Boethian Number Theory
A Translation of the *De Institutione Arithmetica*

Studies in Classical Antiquity
6

Boethian Number Theory

A Translation of the *De Institutione Arithmetica*

(with Introduction and Notes)

by

Michael Masi

Amsterdam - New York, NY 2006

The paper on which this book is printed meets the requirements of "ISO 9706:1994, Information and documentation - Paper for documents - Requirements for permanence".

Transferred to digital printing 2006
ISBN-10: 90-6203-785-2
ISBN-13: 978-90-6203-785-8
©Editions Rodopi B.V., Amsterdam 1983
Printed in The Netherlands

For Susanne,

Mulier et Mater Optima

CONTENTS

Preface

It is only in recent years that the study of medieval science has become a serious enterprise. The production of texts, translations, and surveys will no doubt eventually provide students with more of the necessary tools for an understanding of the sciences and their relationship to other aspects of medieval thought. The editions of the University of Wisconsin in Medieval Sciences have already produced a number of significant and basic books. When a substantial body of works on medieval arithmetic, algorism, geometry and astronomy has appeared, many judgments made concerning the quality and significance of this portion of medieval thought will accordingly have to be revised.

Medieval sciences have traditionally been judged against their Greek or Arabic sources and have generally been found wanting. It is hardly a new claim that these scientific and mathematical works should be evaluated not by Greek or modern standards but in terms of the complex of medieval culture. Ernst Curtius made the point clear in relation to literature in his monumental *European Literature in the Latin Middle Ages*. Unfortunately, the short sighted view of medieval science is still common, particularly in the realm of mathematics. The comment of mathematical historian E.T. Bell "We therefore continue our descent to the nadir of mathematics, and follow the learned Boethius into the abyss"[*] is still cited seriously. As I hope the following pages will demonstrate, Boethius is important and not only for the mathematician. Indeed, in terms of mathematics, he was hardly an innovator.

But he, and the modern historian as well, have been the victims of a persistent anachronism--that mathematicians and scientists be judged by the originality of their contribution. We now understand that Boethius should be studied in terms of those who read his works in the Middle Ages, that is, he should be known by historians of culture and in the total context of medieval ideas.

It is with the hope that all students of medieval culture will find something of importance in the *De Institutione Arithmetica* that this transla-

[*] *The Development of Mathematics* (New York, McGraw-Hill, 1945), p. 88.

tion, with introduction and notes, is presented. I have indicated the possibilities of its study with an introductory survey of the Boethian influence in several arts. The mathematical concepts are basically simple in spite of their complex statements, which are still larded with the Greek syntax and terminology of the works where Boethius read them.

It is to be hoped that there will be further interest in applying his principles to the interpretation of medieval literature and arts.

I owe a debt of gratitude to several institutions and many individuals who helped me in the course of my work on the translation of Boethius. Loyola University has generously provided financial assistance at several stages in my work. In addition to two smaller research grants, the University has made available a $1000 subvention for the publication of this book.The custodians of the Newberry Library, particularly in the Rare Book Room, have been extremely helpful and courteous. Bernard Wilson read and gave helpful comment on the work in an early stage. John Tedeschi also supplied criticism and encouragement when it was much needed. The Institute for Research in the Humanities at the University of Wisconsins enabled me to set aside teaching duties for one year and bring the work to its completion in a peaceful setting near Lake Mendota.

This book is dedicated to my wife who not only encouraged my labors but threw herself with zest into the tasks of proofreading and revision. Professor Edward Lowinsky will always deserve my thanks for continued interest in my Boethian work. John Conley of the University of Illinois, Circle Campus, was also kind enough to read and comment on my translation, as was Domenico Bommarito of Milan, Italy, who spent many patient hours with difficult passages. Others who have given help and shown interest and whom I wish to mention include Richard McKeon, Jerome Taylor, Emmet Bennett, Fannie LeMoine, C. Weinstein, and Loretta Freiling.

Chicago
September 1, 1976

BOETHIAN NUMBER THEORY

Introduction

The influence and popularity of Boethius' *Consolation of Philosophy* are well known to students of the Middle Ages and Rennaissance. But perhaps less well known is the mathematical influence of Boethius, partly because Boethius' mathematical work was seen as only one part of the work of his numerous heirs and commentators and partly because we are less aware of what the study of mathematics and the nature of the *quadrivium* fully involved during the Middle Ages.

The consistency, even into the Renaissance, of the Liberal Arts curriculum,[1] its essentially mathematical nature, its influence beyond the *quadrivium* on music theory and practice, and its bearing on the nature of aesthetics[2] are all revelant to the basic concepts outlined in Boethius' *De Institutione Arithmetica*. Not only does the name of Boethius appear repeatedly in discussions of proportions and harmony, but numerous manuscripts and publications of his works and commentaries on the *De Institutione Arithmetica* continued with undiminished, even increased, vigor into the sixteenth century.

Before I present an outline of this scope of influence, the distinction between practical and theoretical mathematics should be clarified in order to help avoid a common misunderstanding. The modern meaning of *arithmetic* conveys nothing of what it meant for Boethius. The difference between arithemetic ($\dot{\alpha}\rho\iota\theta\mu\eta\tau\iota\kappa\dot{\eta}$) and logistics ($\lambda o\gamma\iota\sigma\tau\iota\kappa\dot{\eta}$) was the same for Boethius as it was for the Greeks who originally defined it.[3] Both disciplines deal with numbers, but arithmetic designates the theory or philosophy of number; only after the Middle Ages did the term designate an elementary discipline of counting and calculation. The process whereby one undertook the solution of practical problems of computation

---

1. *Trivium:* grammar, rhetoric, logic; *Quadrivium:* arithmetic, music, geometry, astronomy.
2. See various chapters in E. de Bruyne *Études d'esthetique médiévale* (Bruges, De Tempel, 1946).
3. See Sir Thomas Heath, *A History of Greek Mathematics* (Oxford, Clarendon Press, 1921), Vol. I, pp. 13-16.

was known to the Greeks and to Boethius as logistics and to the Middle Ages as algorism. [4]

The nature and scope of number theory is adequately explained in the first chapter of the *De Institutione Arithmetica*--it is essentially a preparatory study for philosophy. As such, among the Neo-Pythagoreans, it had a fundamentally moral character and bespoke the order of the world in its most basic terms. The expression of this order was eventually, in the other disciplines of the *quadrivium,* expanded into musical terminology where it acquired the dimension of harmony; in the study of geometry, it was extended to plane surfaces and solid figures. In astronomy, the geometric measurements and the metaphor of harmony found their widest applications in the definition of the order of the universe and in the supreme model of concord, the music of the spheres.

To demonstrate within the limits of this introduction the pervasiveness of Boethius' treatise on the study of number theory, its importance as a preparatory study for music, and the bearing of number theory on architecture, literature, and moral philosophy, I have organized my discussion under five headings. With each I have provided adequate bibliography so that those interested in particular applications of this number theory may pursue and test the application of principles in the *De Institutione Arithmetica* to other disciplines. The five headings are: (1) The Iconography of the Liberal Arts; (2) the *De Institutione Arithmetica* and the *De Institutione Musica* in the theoretical writings of later musicologists; (3) Arithmetic proportion and architecture; (4) Literary extensions of the Theory of Number; (5) Commentaries, derivative studies, and extant manuscripts.

---

4. See Nicomachus of Gerasa, *Introduction to Arithmetic,* trans. Martin Luther D'ooge, intro. Frank E. Robbins and L. C. Karpinski (New York), Macmillan,1926, pp.3-4; Plato, *Gorgias,* Sec 451C; *Theatetus,* Sec. 145A, 198A. For the Middle Ages, see A. C. Crombie, *Medieval and Early Modern Science* (New York, Anchor Books, 1959), Vol. 1, pp. 50-51.

THE ICONOGRAPHY OF THE LIBERAL ARTS AND THE
BOETHIAN ARITHEMETIC

The allegorical portrayal of the seven Liberal Arts presents an irregular tradition.[5] The most obvious discrepancy among the numerous representations of the arts one may find is lack of agreement among the artists concerning the proper order of disciplines. Proper order is significant since in the first chapter of the *De Institutione Arithmetica,* Boethius postulates a necessary logical sequence in the study of the quadrivial disciplines: arithmetic, music, geometry, astronomy. We may come to terms with this disparity in the iconographic material if we realize the persistence of two basic and differing representational traditions for the Liberal Arts, one stemming from the *Liber De Nuptiis Mercurii et Philologiae* of Martianus Capella and the other from Boethius. Capella's treatise, written in the fifth century, was well-known throughout the Latin Middle Ages and made a very significant contribution to the study of philosophy and the arts. In it, allegorical figures of the Liberal Arts are described in a highly pictorical manner and they appear in this order: geometry, arithmetic, astronomy, music. The essential difference between this order of the *quadrivium* and the sequence of disciplines demanded by Boethius depends on whether music is considered a mathematical study and is paired off with arithmetic (Boethius) or an harmonic study and paired off with astronomy (Capella).

5. There has been a number of studies on the inconography of the Liberal Arts. These include: J. A. Clerval, *L'enseignement des arts libéraux à Chartres et à Paris* (Paris, A. Picard et Fils, 1880); E.Mâle, »Les arts libéraux dans la statuaire des moyen-âge,« *Revue Archéologique,* Vol.17 (1891), pp. 339 ff. P. D'Ancona, »Le rappresentazioni allegoriche delle arti liberali nel Medievo et nel Renascimento,« *L'Arte,* Vol. 5 (1902), pp. 137-55, 211-28, 269-89, 370-82; K. Kunstle, *Iconographie der Christlichen Kunst,* Vol. 1 (Freiburg, Herder, 1928), pp. 145-56; Adolf Katzenellenbogen, »The Representation of the Liberal Arts,« *Twelfth Century Europe and the Foundation of the Modern Society,* edds. M.Clagett, Gaines Post, R. Reynolds (Madison, University of Wisconsins Press, 1961), pp. 39-55; Phillippe Verdier, »L'iconographie des Arts libéraux dans l'art du moyen-âge jusqu'à fin du quinzieme siècle,« *Arts libéraux et philosophie au moyen-âge* (Montréal, Institut d'études médiévale,1969), pp. 305-55; Michael Masi, »A Newberry Diagram of the Liberal Arts,« *Gesta,* 11/2 (1973), 52-56; Michael Masi, »Boethius and the Iconography of the Liberal Arts,« *Latomus,* 33 (1974), 57-75.

We should note that the treatise of Capella did not provide the essential texts which came to be studied for the Liberal Arts in the later medieval schools and universities. The compendium of disciplines by Capella is too brief; it was the Boethian texts of arithmetic, music, and geometry (by attribution) which certainly by the later Middle Ages became almost standard as the text books for those disciplines. As a result of the order in the *De Nuptiis,* however, and probably because Martianus Capella provided the graphic descriptions, an iconographic tradition developed which confused the notion of order among the disciplines, if the notion of a proper order even occurred to the artists. In many representations music is last, after astronomy. When this is the case, we may assume some remote influence of Martianus Capella or of one of his commentators. When we find music in the first or second place paired with arithmetic, we may assume some adherence to the Boethian tradition.

While there were two basic orders known in the iconographic tradition, the manuscript evidence indicates that students in schools and universities studied these disciplines in the Boethian order. Where two or more treatises of the *quadrivium* are now found bound together, these are almost always in an order which testifies to the Boethian sequence. Such texts provide a testimony more reliable than iconography for determining the order followed in the curricula of the colleges.[6]

Another aspect of the allegorical portraits is the representation of the »master« or author of the text for each discipline. The arrangement of masters and allegorical figures may be clearly understood in a unique drawing from a twelfth century manuscript of the Boethian *De Institutione Musica* in Chicago,[7] seen in figure 1. Here a series of fourteen niches surrounds a central circular area, and the total scheme resembles a rose window or wheel. The written descriptions in the various compartments of the scheme define the figures which portray, alternately, the allegorical figures of each discipline (the female allegorical figure is technically designated a *virgo)* and a master traditionally assigned as an ancient authority on the sub-

6. Examples of the *De Institutione Arithmetica* bound with the *De Institutione Musica* in the Boethian order may be seen in these manuscripts: Cambridge University Library, Ms Ii. III, 12, S. XI; Oxford, Balliol College, Ms 306, S. X-XI. Trinity College, S. X; Vienna, Österreichische Nationalbibliothek, Ms 55, S. X; Ms 2269, S. XI. A manuscript at Prague, Universitnl Knihovna, Ms 1717, S.IX, contains the *De Institutione Arithmetica De Institutione Musica, De Geometria* in that order. Capella's placing of music in final position derives, no doubt, from Plato. See *Republic,* Book 7, sec. 530-531.
7. Newberry Library, Ms f. 9, S. XII, f. 65[V].

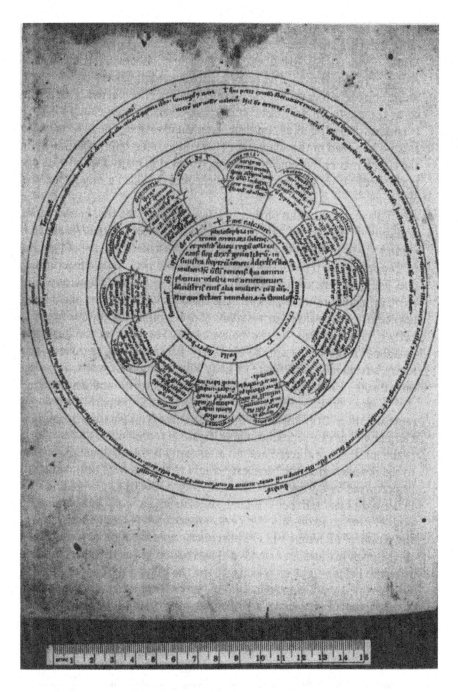

Figure 1: Newberry Library, Ms f. 9, f. 65ᵛ.

ject or, sometimes, the author of the standard text. Beginning at the upper right, these are Dialectic (Aristotle), Rhetoric (Tully), Grammar (Priscian), Music (Pythagoras), Arithmetic (Nicomachus), Geometry (Euclid), and Astronomy (Ptolemy). The order is fundamentally Boethian, but with music coming before arithmetic. In this diagram, Nicomachus is represented as the master of arithmetic study. His Greek treatise could not have been known well, if at all, until the late Middle Ages. The Nicomachean treatise was read in the adaptation of Boethius, who refers continually to the Greek master. Nicomachus was represented in many Liberal Arts diagrams because his greater antiquity no doubt provided a more venerable authority.[8]

This diagram has an obvious relationship with the well-known representation of the Liberal Arts by Herrad of Landsberg in the *Hortus Deliciarum* (f.32). In this late twelfth century representation, none of the masters appears, but the circular diagram (figure 2) with the architectural niches has some unknown relationships to a bronze engraved bowl as well as to other manuscript illustrations. The maidens who represent the disciplines seem to be engaged in a stately dance around the central figure of Lady Philosophy and the two philosophers. The order of the quadrivial disciplines is the same as in the Newberry scheme (music, arithmetic, geometry, astronomy). The figure of Lady Philosophy is crowned and seated on a throne. From her sides flow seven streams which nourish the arts.

A series of drawings from Munich certainly forms part of a group of representations from the early thirteenth century originating in southern Germany. The one reproduced here (figure 3) represents the double niche of the master of arithmetic and his attendant allegorical figure. According to the description in the Newberry diagram, the figure of arithmetic is to be portrayed by a *Virgo in destera tenens virgulam* (A virgin holding a rod in her right hand.). In her left hand, she is to hold a scroll with this inscription: *Per me cunctorum scitur virtus numerorum* (Through me is known the power of all numbers). The allegorical representation of arithmetic in the Munich manuscript extends a flowered rod in her right hand and her left hand holds a scroll on which is found the identical inscription; yet here Boethius replaces Nicomachus and thereby takes his rightful place adjoining the figure of arithmetic. This portrait is remarkable in that it represents

---

8. For a description of this diagram with transcription and translation of the Latin comments and some observations on its relationship to Herrad's diagram, see my article »A Newberry Diagram of the Liberal Arts,« cited above.

Figure 2: Herrad of Landsberg *Hortus Deliciarum*, f. 32

Figure 3: Munich, Staatsbibliothek, Ms 2599, f. 102ᵛ.

a majestic and dignified master and, in a series of portraits where the artist has meticulously followed every letter of the Newberry instructions, Nicomachus is deliberately changed to Boethius. It is a Boethius of noble physical deportment and as spiritually imposing as the allegory. The ancient authors for the texts of the Liberal Arts were often suffered to squat undignifiedly subservient to the allegory of their respective disciplines. The best examples of subservient authors, servants more than masters of their disciplines, may be seen around the Incarnation Tympanum of the Royal Portal at Chartres.[9] But the artist of the Munich manuscript has brought Boethius to his full dimensions as author of the treatise; not only does the artist grant him his rightful place in a position usually assigned to another writer, but, in addition, extends to both figures a scroll whose statement indicates the source of the power of number. The theory of number issues equally from the discipline and from the author of the work.

Other iconographic representations manifest various attitudes toward the work and its author. One interesting representation in the House of the Apostelkloster in Cologne from the twelfth century[10] seems to acknowledge the double tradition of the allegorical representations. In it Boethius and Martianus Capella appear seated together.

A final example has been chosen as particularly meaningful since it appears in the early Renaissance and demonstrates a persistent vitality in the tradition. No one interested in the iconography of the Liberal Arts can overlook a series of woodcuts printed in a compendium of learning assembled by Gregorius Reisch entitled *Margarita Philosophica*. This work first appeared in Freiburg, 1503, and its twelve or more subsequent editions evince great popularity. Many of the later versions expanded elaborately on the contents of the original, and an Italian translation appeared in 1594 and 1599. The general structure, however, and the illustrations of the original work were maintained. It contained chapters on each of the seven Liberal Arts, on Moral Philosophy, on Metaphysics, and on Theology. With some of the chapters there appeared a woodcut allegorically representing an aspect of each study. Two of these may be singled out.

The *Typus Grammaticae,* which accompanies the text of the first discipline, represents a virgo of great stature; in one hand she extends a table with the alphabet to an approaching student and with the other she inserts

9. See Katzenellenbogen, »Liberal Arts,« figure 2.
10. Paul Clemen, *Die Romanische Monumentalmalerei in den Rheinlanden* (Düsseldorf, Schann, 1916), p. xxxv.

Figure 4: *Margarita Philosophica*, Typus Grammaticae

Figure 5: *Margarita Philosophica*, Typus Arithmeticae

a key to open the door opening on the first level of a structure which suggests the university (figure 4). In a room on the lowest level of the stucture are Priscian and Donatus, both known as authors of grammatical texts. (It seems appropriate that two masters be assigned here.) The second level represents small figures peering from windows labeled Aristotle (Logic), Tully (Rhetoric), Boethius (Arithmetic). The third level represents Pythagoras (Music), Euclid (Geometry) and Ptolemy (Astronomy). Above them, on the fourth level, two faces bear titles for the two aspects of medieval philosophy: *philosophia practica* and *philosophia theoretica*. From the top, which is open to the skies, Peter Lombard, representative of theological studies, surmounts the entire scheme. It is quite likely that the artist had in mind a representation of Lady Philosophy from the *Consolation* where her head is described as reaching to the heavens and her robe as having inscribed on it the letters π *(practica) and θ (theoretica)*.

The *Typus Arithmeticae* (figure 5) bespeaks an intense relationship of Boethius with the Liberal Arts as writer of the text and representative or heir of Pythagoras. In this woodcut, a virgo holds scrolls identifying two men, clad in the garb of sixteenth century northern Europeans, as Boethius and Pythagoras. Pythagoras, the older man, is making calculations on an abacus. Boethius, the younger, makes his computations by a more modern method but maintains a fixed gaze on the computations of the older man. On the left thigh of the figure behind are the numbers 2,4,8 (doubles of each other) and on the right thigh 3,9,27, (triples of each other); these are series which form the basis of many discussions in the Boethian *De Institutione Arithmetica*.

The Iconography of the Liberal Arts represents a large area to be investigated by the historians of both art and philosophy. In its development, there is a wedding of textual and pictorical traditions. Existing studies of this material are, for the most part, elementary surveys of the pictorical traditions; a book needs to be written which encompasses the two traditions. In that book, the part of Boethius should be seen as increasing in significance during the late Middle Ages; this significance, which begins in the texts, will become more evident in the pictorical representations.

BOETHIAN NUMBER THEORY AND MUSIC

As is readily apparent from the testimony of both iconography and manuscript inventories, the *De Institutione Arithmetica* is an introduction to the principles of music within the structure of disciplines of the *quadrivium*, and knowledge of its contents is essential for an understanding of music theory. In the work on arithmetic, we find definitions of terms and of proportional principles used to determine the pitches of the musical scale; these are the terms in which the Boethian explanations of harmonies and dissonances must be understood. In the treatise on number theory, Boethius provides the elaborate terminology of proportions which he only briefly reviews in its musical applications in the second treatise. [11]

Consequently, any understanding of the importance and the popularity of the Boethian *De Institutione Musica* can illuminate the immediate and very significant extension of number theory into the beginnings of musical art forms. Such a study provides testimony for the importance of Boethius in the study of medieval and Renaissance musical theory.

An examination of music, as it depends on Boethian number theory, must be made through the proportional principles. The derivation of music theory from arithmetical proportionality may be seen readily in the works of the following authors. This list contains the most significant writings (based on the Boethian formulation of number theory) which were commonly used by composers and music theoreticans. From these works we may proceed to practical musical applications.

1) Jacques de Liège, *Speculum Musicae,* [12] E. de Coussemaker, ed. *Scriptorum de Musica Medii Aevi* (1864-76), Vol. 2, pp. 196-213. This

---

11. The precise relationship between the two treatises is discussed at length in a forthcoming article by Ubaldo Pizzani. Pizzani challenges the notion that by the time Boethius wrote the *De Institutione Musica* he still intended the relationship between the disciplines as outlinded in the proemium to the treatise on arithmetic. His arguments are compelling but by the twelfth century it cannot be doubted that the schema of the *De Institutione Arithmetica* prevailed. It is my feeling that regardless of Boethius' intention by the time he wrote on music, his readers eventually took his earlier relationships among the disciplines of the *quadrvium* as a useful way of organizing their study.

12. For a discussion of the authorship, meaning, and influence of this work, see F. Joseph Smith, »A Mediaeval Philosophy of Number,« *Arts libéraux et philosophie au moyen-âge* (Montreal, Institut d'études médievale, 1969), pp. 1024-39.

work represents a thorough integration of number theory with philosophical and theological speculation.

2) Johannes de Muris, *Libellus Cantus Mensurabilis.* (Cous. Vol. 3, pp. 58-79). De Muris presents a rudimentary application of Boethian proportion to musical writing in a direct and practical manner. His discussion centers mainly on the *Dupla in diminutione.*

3) Guilelmus Monachus, *De Preceptis Artis Musicae Libellus* (Cous. Vol. 3, pp. 277-88). This *Libellus* is the first truly significant discussion of the system of proportional writing (c. 1460), and its exposition provides many examples. It remained, for a long time, a basic treatment on the matter.

4) Tinctoris, *Proportionale Musicae* (Cous. Vol. 4, pp. 153-77).

5) Gaffurio, *Practica Musica,* Ed. and trans. Clement A. Miller (Dallas, American Institute of Musicology, 1968). This highly elaborate treatise carries proportional discussion much farther than any of its predecessors and spins out most thinly all the theoretical possibilities. Contrary to the emphasis in the title, it is not concerned with the practical aspects of proportion. Gaffurio is instead concerned with working out the details of such remote proportional diminutions as the ratio of 9:23.

6) Prosdocimus de Beldamandis, *Brevis Summula Proportionum* (Cous. Vol. 3, pp. 258-428). This work was long a standard study for proportions and, as we shall see, is a handy reference for an outline of the five proportional categories.

7) Sylvestro Di Gnassi, *Opera* (Venice, 1535). This was an important and standard source.

In addition to these writers, there were many minor medieval and Renaissance theoreticians who, regardless of their influence, certainly gave testimony to the importance of the Boethian treatise by their incorporation of his basic principles into their own works. For a sample of these, one may examine the collection of treatises easily available in the edition by Coussemaker.

Vol. 1, Jerome de Moravia, *Tractatus de Musica,* pp. 3-4, 6-7, 10-16,35-70;
------ Petrus Picardi, *Musica Mensurabilis,* pp.139-52.
------ Walter Adington, *De Speculatione Musicae,* pp. 192, 194, 205,211;
------ Quidam Aristoteles, *Tractatus de Musica,* p. 252;
Vol. 2, Dominus Guidonus, *Regulae de Arte Musica,* p. 129;
Vol. 4, Simon Tunstede, *Quattuor Principalia Musicae,* pp. 201-7.
------ Johannes Gallia, *Ritus Canendi Vettissimus et Novus,* pp. 298-349.[13]

---

13. One may refer as well to a collection of early treatises edited by Martin Gerbert, *Scripto-*

The musical proportion most often found is that most easily perceptible to the ear, the proportional change of tempo. A change of tempo may
be found in sophisticated and popular music of all times and is hardly a
Boethian function by itself. In its simplest form, it consists of doubling or
halving the rhythmic tempo. When music is played *alla breve,* for example, the value of the notes is diminished by half and the music's speed is
doubled. This is a common form of tempo proportioning and when it
occurs the tempo has changed in a ratio of 2:1 or a duple.

However, during the late Middle Ages and during much of the Renaissance, this simple mode of tempo change was only the beginning of
proportional changes far more elaborate. Such proportionalities were first
described by theoretical writers, then were adopted, though only to a limited extent, by the composers. We may draw illustrative material from the
work of Sylvestro di Gnassi, who provided theoretical writings and practical examples.[14]

Sylvestro carefully defines a system of diminutions or tempo reduc-
tions. The first rule he describes is for the dissolution of a semibreve into
four semiminums (a whole note into four quarter notes: o: ♩♩♩♩). This reduction translates arithmetically as a multiplex ratio of four to one. The
second rule of Sylvestro instructs the composer in the dissolution of the
semibreve into five semiminums; the third Rule, into six; the fourth, into
seven. He then provides for changes of tempo from four semiminums into
seven (♩♩♩♩: ♩♩♩♩♩♩♩). This is a supertripartient ratio, that is one in
which the smaller number fits into the larger with the remainder of three.
The change of tempo from five to six is a sesquiquinta, that is from a smaller number into a larger number where the smaller fits into the larger with
a remainder that is one fifth of the smaller. The composer is also instructed
in the changes of the superbipartient (five and seven) and the sesquisexta
(six and seven).

These changes can hardly be detected, let alone identified, by the ear
of even the most careful listener. Such musical composition makes it clear
that proportional writing is often not a matter of music but of a higher understanding derived from music. (Curt Sachs does not do it justice when

res Ecclesiastici de Musica Potissimum, 3 vols. (San Blasianis, Typis San-Blasianis, 1784; reprinted, University of Rochester Press, 1955) and see there further Boethian references. For
manuscript evidence, see my article »Manuscripts Containing the *De Musica* of Boethius,«
Manuscripta, 15 (1970), 89-95.
14. See Imogene Horsley, »Improvised Embellishment,« *Journal of the American Musicological society,* 4 (1951), 7.

he describes it merely as a matter of »pen and ink«.)[15] In such cases, the presence of proportional architecture in the tempo changes is the composer's tribute to a notion of beauty and proper order which transcends the ability of the senses alone to perceive it and which, therefore, appeals primarily to the intellect which, upon reflection and analysis, is able to grasp its real meaning. This proportional writing ideally lent to music a beauty of order which arose from elements not necessary for its musical structure; it was an order which spoke to a higher notion of propriety and which emerged from the creator's desire to make the work of art conform to an absolute concept whose nature was derived from the structure of the universe and, ultimately, as Boethius puts it, from the mind of God.

The complete outline of tempo proportions which became more or less standardly known during the Renaissance was provided by Prosdocimus de Beldamandis (1370?-1428). Prosdocimus was thoroughly familiar with the Boethian sources for his material and as well as a musical treatise, he wrote a study on mathematics in which he acknowledged his debt to »Bohectius« for the definition and elaboration of number theory.[16]The following table is adapted from his *Brevis Summula Proportionum:*

I. *Multiplex proportion,* in which tempo is:
a) Doubled.
b) Tripled.
c) Quadrupled.
II. *Superparticular proportion,* in which the faster exceeds standard tempo by some aliquot part:[17]

15. On the nature and use of such changes, see his *Rhythm and Tempo: A Study in Music History* (New York, W.W. Norton, 1953), pp. 206-214.
16. See David Eugene Smith, *Rara Arithmetica* (Boston, Ginn and Company, 1908), p. 13. The *Brevis Summula* may be found in Coussemaker, Vol. 3, pp. 258-61. Compare this material with Boethius, *De Institutione Musica,* Book 1, chap.23-31.
    Boethius and those who adopt his classification of proportion do not distinguish between *ratio* and *proportio.* According to Euclid, however, a ratio is a relation between magnitudes of the same kind while a proportion is a relation between ratios. Thus, »a proportion in three terms is the least possible« *(Elements,* Book 5, chap. 8.). In the *De Institutione Arithmetica,* Boethius uses *proportio* to refer to relations of two as well as of more terms.
17. An aliquot part divides into a whole without a fractional remainder. 1 is such a part of any integer. In multiplex proportions, 2:1 or 3:1, the faster tempo, since it is a simple multiple of a standard, can always be divided by the standard without a remainder. The same is true of the superparticular. The excess of 1, as in 3:2, is necessarily an aliquot part of standard. (Cf. Sachs, p. 206). In a sesquialter, the remainder is half the smaller number; in sesquitertial, it is a third; in sesquiquartal, a fourth, etc.

Figure 6.

a) Sesquialter, 3:2.

b) Sesquitertial, 4:3.

c) Sesquiquartal, 5:4.

III. *Superpartient proportions,* in which the faster tempo exceeds standard tempo by an aliquant or non-aliquot part, as:

a) Superpartient, 5:3 or 7:5 or 9:7.

b) Supertripartient, 7:4 or 8:5 or 10:7.

c) Superquadripartient, 9:5 or 11:7 or 13:9.

The final two categories contain proportions in which the faster tempo holds the aliquot or non-aliquot part in addition to at least twice the standard tempo:

IV. *Multiplex superparticular proportions,* as the duplex superparticular, 7:3.

V. *Multiplex superpartient proportions,* as the duplex superbipartient, 8:3.

Although tempo proportions were the most common and most easily understood musical adaptations of proportional number theory and were worked out to the finest detail by writers such as Gaffurio, proportional arrangements of other kinds were also used. Lengths of movements, the number of notes in a melody or a measure, relative numbers of measures in various parts of a single or related pieces were governed by the same proportional principles that we noted in tempo structuring. A simple but helpful analysis of the use of such proportional writing may be found in the development of the *cantus firmus* through the parts of the *Missa Se la face ay pale* by Guillaume Dufay.[18]

The tenor of the mass provides the essential thematic material for all movements and defines the fundamental unity for the structure of each part; it affords Dufay numerous possibilities of proportional variations in diverse rhythmic combinations with other voices and with variations of its own tempo changes. The parts of the mass may then be analyzed in terms of the *cantus firmus* in figure 6:

1) KYRIE. In the first movement, each note of parts A and B is augmented to twice its original value for the initial triple Kyrie, but no significant variation occurs in the Christe. In the final triple Kyrie, the C part is aug-

18. This analysis is based on the studies of Gustav Reese, *Music in the Renaissance* (New York, W.W. Norton, 1959), pp. 69-71; C. Van den Borren, *Guillaume Dufay: son importance dans l'évolution de la musique au XVe siècle* (Bruxelles, Maurice Lamertin, 1926), pp. 104-127; P. Wagner, *Geschichte der Messe,* I Teil, bis 1600 (Leipzig, Breitkopf & Hartel, 1913), pp. 89-100.

mented. The result is the correspondence in one measure of tenor melody to two measures of the other voices whenever augmentation occurs.

2) GLORIA. In the lengthy second movement, Dufay maintains this basic three-fold structure. The first part includes the music for the text: *Et in terra pax hominibus bonae voluntatis. Laudamus te. Benedicimus te. Adoramus te. Glorificamus te. Gratias agimus tibi propter magnam gloriam tuam. Domine Deus, Agnus Dei, Filius Patris.* In the music for this text, each measure of the tenor corresponds to three measures of the other voice parts. The tempo designation remains, as in the original statement, and as in the Kyrie -- 3/4 or triple meter. The score for the second part, however, makes a change to double meter. The music for this part accompanies the text: *Qui tollis peccata mundi, miserere nobis. Qui tollis peccata mundi, suscipe deprecationem nostram. Qui sedes ad dexteram Patris, miserere nobis. Quoniam tu solus sanctus. Tu solus Dominus. Tu solus altissimus, Jesu Christe.* In this section of the composition, the tenor is still in triple meter, or 3/4, but all the other parts are in duplex, or 2/8. Each measure of the tenor melody still corresponds to three measures in the other parts.

In the final passage, *Cum Sancto Spiritu, in gloria Dei Patris, Amen,* there is a complete return to original time values. All the voices are in 3/4 tempo, and the tenor is in its original form so that the correspondence between each measure of the tenor and the other voices is one to one. The parts of the Gloria, then, traverse an obvious and very Boethian scheme of 3:2:1.

3) CREDO. Here the same general arrangement prevails as in the *Gloria.*

4) SANCTUS. In this movement, the first musical division includes the text: *Sanctus, Sanctus, Sanctus, Dominus Deus Sabaoth.* In it there appears only segment A of the tenor and it is written against the other voices in augmented time values, 2:1 so that each measure of the tenor corresponds to two of the other voices, as in the Kyrie. For the text *Pleni sunt caeli et terra gloria tua,* the tenor is silent, but it appears again, with double time values, using the B segment of the melody, for the text *Hosanna in excelsis.* In the music for *Benedictus qui venit in Nomine Domini,* the tenor again remains silent, thereby interposing two silent parts between three segments of the *cantus firmus.* The final section, *Osanna in excelsis,* contains only the C portion in double time, but for this last segment, Dufay has put the other voices in double meter as well.

5) AGNUS DEI. The last movement, with its naturally triplicate text, follows the musical structure of the Kyrie and so establishes a return; the first and last sections join a musical cycle.

Certainly such musical architecture is not perceptible to the listener.

These larger musical plans are discovered on analysis and as such, they accord with the larger ideal order to which literary and other artistic structures speak in similar terms.[19]

---

19. For another and more elaborate analysis, see M. Van Crevel, ed., Jacob Obrecht, *Missa Maria Zart, Missae VII* (Amsterdam, G. Alsbach, 1960).

ARITHMETIC PROPORTION AND THE MEDIEVAL CATHEDRAL

Otto Von Simson, Paul Frankl, H. Beseler, H. Roggenkamp,[20] and others have made various connections between medieval cathedral design and Boethius, each for his own interesting purposes in the explanation of medieval architecture. However, there is a theory of beauty and order derived from the study of these cathedrals which may be related to the express function of the medieval educational curriculum, through the sequence of the Liberal Arts. This concept of beauty is correlative with the notion of proper moral order. In my survey of the medieval cathedrals, necessarily brief and superficial, I will attempt to outline some basic relationships between the structural design of these buildings and other material discussed in this introduction. In the process, I will draw together ideas from analytical and historical studies mentioned in the notes and show how the Boethian *De Institutione Arithmetica* and, to a certain extent, the *De Institutione Musica,* are often directly connected with the intellectual milieu out of which the concepts of cathedral design emerged.

Though it is not the first, certainly the most significant school for Boethian mathematical studies, which is also connected with an important medieval cathedral, is Chartres. Other structures may be more convincingly submitted to a rigorous Boethian analysis, but the cathedral of Chartres in many ways does justice to the notion of order and proportion that Boethius demanded of man's attitude toward the world. Von Simson has demonstrated some of the proportionality of Chartres' design and attributed its planning to concepts emerging from the Boethian and Augustinian treatises on music, both of which specify similar laws of aesthetics that depend on proportion and measure. But the laws of proportion which von Simson invokes should be referred more properly to the principles of Boethius' *De Institutione Arithmetica.*

The purpose of proportional structure in the cathedrals is identical to its purpose in other works of art, that is, it supplies the principles which

20. Otto von Simson, *The Gothic Cathedral* (New York, Harper and Row, 1956); H. Beseler and H. Roggenkamp, *Die Michaeliskirche in Hildesheim* (Berlin, G. Mann, 1954); Paul Frankl, *The Gothic: Literary Sources and Interpretations* (Princeton, Princeton University Press, 1960).

raise the elements of design to the highest levels of understanding. True proportionality, according to the principles of the *De Institutione Arithmetica,* lifts the physical design of a structure out of the realm of mechanical and engineering structural requirements to a level of completeness and abstract suitability accountable only to the mind. The ideal structure of the cathedrals responds to a superior law of order, one linked with the metaphor of the harmony of the universe and expressive of the moral law manifested in God's creation. It is the same order arrived at by the student's matured understanding achieved upon completion of the curriculum of the seven Liberal Arts, and it corresponds to the order defined by the discipline of moral philosophy and articulated in the ascending mathematical disciplines in the *quadrivium.* In several ways, the cathedrals, whether in the design of the entire structure or in only part of that structure, become a tribute to the ongoing study of such arts and manifest their fruition in the moral order.[21]

The Cathedral school of Chartres nurtured a vigorous revival of Pythagorean and Platonic studies which, under the particular inspiration of Boethius and during the time of the learned bishops Fulbert, Gosselin of Mussy, Robert le Breton, William, and, culminating with John of Salisbury (1176-1178), was particularly fruitful in terms of Christian theological writings. During the eleventh and twelfth centuries, the scholars of Chartres exhibited great enthusiasm for mathematical pursuits and produced commentaries on the *quadrivium* and on the works of Boethius that still stand out from the entire period of medieval philosophy. Foremost among these scholars were Thierry, his brother Bernard, and their teacher, Gilbert de la Porrée. The *Heptateuchon* of Thierry, a compilation of texts for the study of the Liberal Art, prominently featured the work of Boethius.[22]

For the schoolmen of Chartres, mathematics was the link between God and the world, the intellectual key which unlocked the secrets of the

---

21. Professor Lon Shelby has made several valuable contributions to the understanding of practical mathematics and medieval cathedral structure. (See his articles in the bibliography.) It is impossible to make an adequate response to his case against the Boethian interpretation of cathedral design in the space available here. I am not convinced, however, that he has refuted the essential validity of von Simson's argument.

22. See B. Haureau, *Notes et extraits de quelques manuscrits Latins de la Bibliothèque Nationale* (Paris, 1890), p. 64 ff; W. Jansen, »Der Kommentar des Clarenbaldus von Arras zu Boethius *De Trinitate*,« *Breslauer Studien zu historischen Theologie,* 8 (1926), pp. 122, 62, 108, 125; E.Gilson, *History of Christian Philosophy in the Middle Ages* (New York, Random House, 1955), pp. 139-53, 619-25.

universe. Thierry extended mathematical studies beyond the Greek logical and ethical preoccupations reflected in Boethius to the realm of Christian theological inquiry. With the help of arithmetic and geometry, he attempted to discern and explain the workings of God in all of creation. The Trinity, with its rich implications of mathematical possibilities, was an obvious field of speculative spade work for him, and Thierry attempted to explain its nature in terms borrowed from the treatises of the *quadrivium*. The equality of the three persons is represented, according to him, by the equilateral triangle. The square of the sides unfolds the ineffable relationship between the Father and the Son.[23]

Von Simson's musical analysis of the Chartres cathedral is valid in its own way and demonstrates how a building can be described as »frozen music« in a manner that the poetical mind which coined the phrase did not fully grasp. Yet von Simson's demonstration of musical proportionality can distract us from a fundamental tie to a basic mathematical proportionality. What the von Simson musical demonstration really comes to is an application of mathematical proportions to cathedral design. The building's basic ground plan, as von Simson describes it (pp. 207-8), is the pentagon and from it are derived the proportions of the Golden Mean [24] for the elaboration of the other proportions of the building. The relationship between the height of the shafts (13.85m.) and the distance between the base of the shafts and the lower string-course (5.35), and the height of the piers (8.61) is in this proportion.

The diagrams and descriptive comments in the sketchbook of Villard de Honnecourt provide examples of geometrical and arithmetic proportionality which begin with the human figure and extend to a model of the perfect cathedral. Villard was closely connected with Chartres and very possibly studied with one of the masters of that school. Since Villard's comments and diagrams indicate a close relationship between the human form and cathedral structure, they make explicit the mathematican's search for the perfect natural model from which may be derived mathematical definitions for the ideal structure. Villard provides the plan for a

23. See J. M. Parent, *La Doctrine de la création dans l'école de Chartres* (Paris, J. Vrin, 1938), p. lll; John E. Murdoch, »Mathesis in philosophiam scholasticam introducta,« *Arts libéraux et philosophie*, pp. 215-55.
24. The Golden Mean is a proportion of three measurements in which the smaller is to the larger as the larger is to the total of the two. Its description does not occur in the *De Institutione Arithmetica* but may be found, briefly stated, in the *De Geometrica*, Friedlein, ed.,p. 386. See also R.C. Archibald's appendix to Jay Hambidge, *Dynamic Symmetry: The Greek Vase* (New Haven, Yale University Press, 1920), pp. 152-7.

Cistercian church designed *ad quadratum,* that is, one whose proportions are derived from the square which is used to determine the dimensions of the entire structure.[25]

The proportions of Villard's Cistercian church correspond to the Boethian sequence of proportions. Nor were his discussions merely theoretical, since there is evidence that these plans were employed by Cistercians in the construction of their churches. According to Villard's canons, the length of the cathedral's nave is in the ratio of 2:3 to the transept. This relationship may be considered in the proportion of a fifth in musical terms or a sesquialter in mathematical vocabulary. The ratio of 1:2 (duplex), or the octave, occurs between the side aisles and the nave. We find the same relationship between the length and width of the transept and the interior elevation. The ratio of 4:3 of the nave to the choir is a sesquitertial relationship or the musical fourth. The 5:4 relationship of the side aisles taken as a unit and the nave is a third or sesquiquartan. The crossing, liturgically and aesthetically the center of the church, is based on the 1:1 ratio of unison, the mathematical unity, the most perfect of consonaces, and the foundation for all number.

Perhaps the most useful and meticulous mathematical analysis of a medieval church is to be found in Beseler and Roggenkamp's description of the Michaeliskirche in Hildesheim.[26] Their book is particularly relevant for this introduction since it makes clear connections between a text of Boethius' *De Institutione Arithmetica* (still to be seen in the manuscript collection at Hildesheim), the principles of the treatise, the Liberal Arts and their study at Hildesheim. During the tenth and eleventh centuries, Boethian mathematical principles had a definite influence on the thought out of which the design of the cathedral emerged.

Although Michaeliskirche, in spite of its magnificent and complex beauty, plays a role subservient to that of Chartres in the history of medieval architecture, and although Hildesheim was not as significant in the

25. See von Simson, pp. 198-9; Villard's influence on cathedral design is also described in E. Panofsky, »Die Entwicklung der Proportionslehre als Abbild der Stilentwicklung,« *Monatshefte für Kunstwissenschaft,* 14 (1921), pp. 188-219, reprinted in English as »The History of Human Proportions as a Reflection of the History of Styles,« *Meaning in the Visual Arts* (New York, Doubleday and Company, 1955), pp 55-107. See also H. R. Hanloser, *Villard de Honnecourt* (Vienna, 1953).
26. In Beseler and Roggenkamp see especially the chapters »Die Geometrische Deutung,« pp. 127-9; »Das Zahlengerüst,« pp.129-30; »Intellektualle Harmonie,« p. 138-40; »Das Maßgebende Tetraeder,« pp. 140-3. A page of the tenth century *De Institutione Arithmetica,* in a volume entitled »Bernwards *liber mathematicalis*« is reproduced among the plates.

development of scholasticism as Chartres was, yet what went on in the schools and intellectual community there is worthy of some consideration relative to cathedral design. The activity of the cathedral school reached a peak under Bishops Bernward (993-1022) and Godard (1022-1038), approximately a century before the well-known writers of Chartres made their appearance. It was, apparently, Bernward who founded St. Michael as the first Benedictine abbey in Hildesheim (996). This church, with two choirs, two transepts, and six towers, dates from the eleventh century and is a good example of German Romanesque. It, like the Chartres cathedral, is an embodiment of the ideals of the intellectual life that flourished in the community around it.

A rather significant individual emerged from the environs of Hildesheim in the tenth century--Hrosvitha, a writer of Latin plays (c. 935-1000). She entered the Benedictine nunnery at Gandescheim near Hildesheim and remainded there for her entire life.[27] The Latin dramatic writings of Hrosvitha offer some interesting evidence of Boethian presence in Saxony during the tenth century. An awareness of the importance of the mathematical disciplines of the *quadrivium* and of their relationship to philosophy (which we saw in the iconographic materials) found its way into her dramatic presentations in some very literal ways. In her play *Pafnutius*, for example, the hero is negotiating the path to reconciliation with God. In an exchange with his disciples, Pafnutius is asked by one of them that question which must obviously lead to the definition of moral philosophy, »What is music?«

Paf: Disciplina una de philosophiae quadruvio.
Disc: Quid est hoc quod dicis quadruvium?
Paf: Arithmetica, geometrica, musica, astronomia.
Disc: Cur quadruvium?
Paf: Quia, sicut a quadruvio semitae, ita ab uno philosophiae principiae harum disciplinarum prodeunt progressiones rectae.[28]
(Paf: It is a discipline from the quadrivium of philosophy.
Disc: What is this you call the quadrivium?
Paf: Arithmetic, geometry, music, astronomy.
Disc: Why a quadrivium?

27. For the cultural life of that community and information on Hrosvitha, see Otto Beyse, *Hildesheim* (Berlin, Deutscher Kunstverlag, 1926) and Franz Dibelius, *Die Bernwardstür zu Hildesheim* (Straßburg, J.H.E. Heitz, 1907).
28. *Opera of Hrosvitha,* ed. Paulus Winterfeld *(Berlin, Weidmann, 1902), pp. 163-64.* See also the more recent edition of H. Homeyer for its notes, *(Munich, F.Schoningh, 1970).*

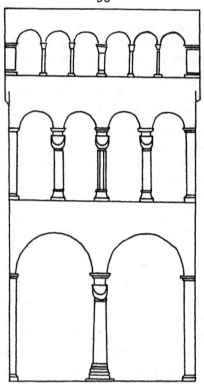

Figure 7a: Gallery of the Angels, Michaeliskirche (adapted from von Simson).

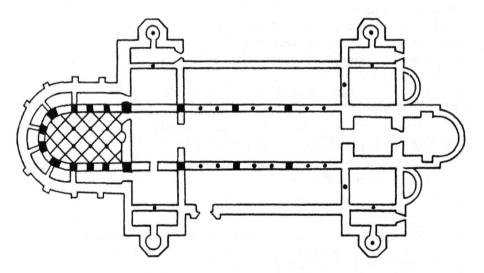

Figure 7b: Outline of floorplan, Michaeliskirche.

Paf: Because as from a path or four-fold way, so from this path and from these principles extend the proper progressions of philosophical principles.)

Again, in another play, *Sapientia*, the main figure is the allegory of wisdom; she engages a disciple in a discussion over diminished, superfluous and perfect number, even and odd number, and other classifications of numbers. The substance of her discourse comes directly from the *De Institutione Arithmetica*, Book 1, chap. 10 (Hrosvitha, *Opera*, pp. 184-6).

The Boethian work on arithmetic was evidently known at Hildesheim. There now exist two copies in the manuscript collection there, and Beseler is justified in making connections between the Boethian number theory and certain aspects of the cathedral's structure. He examines, for instance, the nave and two transepts, each of which describes a series of three large squares (see ground plan, figure 7B) and each of which is in turn composed of three unities of nine feet. The trinitarian symbolism is intentional in a design which interestingly emerges in a pattern of threes and draws measures from the square. Boethian proportions appear in the height of the nave which is twice that of the side aisles and which is in turn equal to the side of the basic squares of the ground plan. In all these relationships we see the presence of the »perfect« sequence of ratios, 1:1, 1:2, 2:3 (a unity, an octave, a fifth). Proportional design is further used at Michaeliskirche in the Gallery of the Angels (figure 7A) where, in the disposition of three levels of arches, the ratios of 1:2:3 are apparent. Note also that the ratios determine as well the width of the supporting columns.

Other church structures may be analyzed in a similar mathematical manner. Examples of Italian churches and their relationship to the design of Villard de Honnecourt appear in the discussions of Paul Frankl (pp. 39-58; 63-70). Of particular interest are the designs for the Milan Cathedral proposed by Stornaloco the mathematician.[29] His severely mathematical design based on the triangle is of theoretical interest only; his plan was never adopted. The failure of such a proportional design in Milan may lead one to speculate on the importance of the community feeling concerning the principles of design. It seems that only communities like Chartres and Hildesheim, where there existed an awareness of mathematical studies and their connection with mathematical works such as that of

---

29. See Camillo Boito, *Il duomo di Milano e i disegni per la sua facciata* (Milan, G. Agnelli, 1889).

Boethius, could a mathematical design be completely realized in the architecture which emerges from the community. Such a unanimity and clarity of feeling apparently did not exist at Milan.

MEDIEVAL LITERATURE AND THE THEORY OF NUMBER

Studies of mystical number theory and the literature of the Middle Ages are well known to students of literary history. Boethius, however, showed no interest in mystical numbers, and though his commentators failed to keep the distinction clear in their writings (see especially Gerardus Ruffus below), his treatment of number adheres conscientiously to a severely philosophical concept. The Boethian preoccupation with number is not part of an interest in the occult; it is an interest in that strictly delimited aspect of number which is clearly comprehensible to the intellect. All the Boethian proportions are based on the intelligible significance of numbers and their calculable combinations. The meanings of numbers come not from secret and hidden associations but from carefully determined relationships, and it is these relationships which delineate the clearly discernible conceptions of beauty in the works of art where we seek them.

It is the function of the sequence in the mathematical disciplines of the *quadrivium* to develop the aspects of proportions extending gradually in widening circles of application to progressively more extensive meanings until the complete mastery of them leads to the comprehension of the moral order and an all-embracive view of the world. The proportions studied in the *De Institutione Musica* are applied to the intervals of sound in the study of music and the proportions are, in connection with that study, invested with the metaphor of harmony. Notions of proportion are applied to planes and solids in the study of geometry; figures extended on planes and into solids are explained and understood in terms of a harmony of parts. This harmony of geometrical parts has obvious applications, as we have seen briefly in Villard de Honnecourt's appeals to geometrical harmony in comparing proportions of human and animal figures to cathedral structure. The student of the *quadrivium* applied geometrical proportions to the movements of the planets in the skies.[30] In astronomy, harmo-

---

30. Boethius may have written a treatise on astronomy, but there is none extant which has been attributed to him. His treatise on geometry was known in the early Middle Ages, but it is now clear that the two geometry treatises extant from the Middle Ages which bear his name are not from his own hand, though they are in some way derived from his work. See Menso Folkerts, *»Boethius« Geometry II: Ein Mathematisches Lehrbuch Des Mittelalters* (Wiesbaden, Franz Steiner Verlag, 1970).

ny and proportion and the music of the spheres present an embodiment of the intellectual ideals studied in the sequence of the Liberal Arts. From that point, as the iconography of the Liberal Arts consistently demonstrates, the next step is to Lady Philosophy, into the realm of the Boethian *Consolation of Philosophy.* This is a transition made clear by the inscription in the central circle of the Newberry Diagram on a scroll held by a maiden accompanying Lady Philosophy: "Qui contemplantur celestia, me venerantur"(Those who contemplate heavenly things venerate me) (figure 1).

Instead of counting lines and totaling up cabalistic number references, I propose to follow a different path from the study of number theory into the medieval uses of mathematics in literature. Literature and numbers may be comprehended within a larger scope, one which extends from the study of the *quadrivium* and includes an ethical context. Ethics is a philosophical discipline but its nature in many medieval works of literature is conditioned by the study of the mathematical sciences of the Liberal Arts. Accordingly, metaphorical references to music, geometry, and astronomy are based on elementary mathematical concepts derived from the study of the *De Institutione Arithmetica.*

The ethical implications of the study of the quadrivium become apparent from the beginning of the *De Institutione Arithmetica.* The *quadrivium* is a four-fold path to the study of the moral truths of the *Consolation,* and Boethius is insistent that these steps be taken carefully and methodically. Nor do the ethical meanings of mathematical study disappear from the treatise after the well known moral statements of the first chapter. These moral reverberations are consistently felt in explications and references throughout the treatise. In chapter nineteen of Book 1, for example, Boethius discusses perfect numbers, that is numbers equal to the sum of all their possible dividends, e.g. $1 + 2 + 3 = 6; 1 + 2 + 4 + 7 + 14 = 28$. These he compares to virtues and vices. As with perfect numbers, virtue is rare, and most moral behavior is short or in excess of the virtuous middle. I have already pointed out several moral associations in Boethius' text and referred to Aristotle and others who have taken note of similar moral applications in discussions of number theory. The transition from arithmetic to music intensifies the moral preoccupations. (*Musica vero non modo speculationi, verum etiam moralitati conjuncta sit.* Music is related not only to speculation, but is indeed closely connected to morality. [*De Institutione Musica,* Book I, chap. 1.])

The concept of world order expressed in the metaphor of harmony manifests an extension of proper proportion to all the elements in the universe; harmony exists between the spheres of the heavens, as it should be found among the nations of the world, among individuals, between man and wife, among a man's parts, both physical (arms, legs, eyes, etc.) and spiritual (intellect and will, reason and passion, etc). Of course, in the domain of human behavior, there is occasionally a lapse in proportion. The ideal of proper order in the spheres and in the elements of the universe is the measure for man's adherence to the moral law, yet man does sometimes fail to meet that measure.

This ethical extension of order and proportion is found consistently in the works of Boethius and is beautifully expressed in the eighth poem of Book 2 of the *Consolation:*

> That the universe carries out its changing process in concord and with stable faith, that the conflicting seeds of things are held by everlasting law, that Phoebus in his golden chariot brings in the shining day, that the night, led by Hesperus, is ruled by Phoebe, that the greedy sea holds back his waves within lawful bounds, for they are not permitted to push back the unsettled earth--all this harmonious order of things is achieved by love which rules the earth and the seas, and commands the heavens.

> But if love should slack the reins, all that is now joined in mutual love would wage continual war, and strive to tear apart the world which is now sustained in friendly concord by beautiful motion.

> Love binds together people joined by a sacred bond; love binds sacred marriages by chaste affections; love makes the laws which join true friends. O how happy the human race would be, if that love which rules the heavens ruled also your souls.[31]

This is the ultimate object of the Boethian number theory as he sees it. From the discipline of the Liberal Arts, one rises to the higher understanding of the nature of things. As the planets, the seasons, the four elements, night and day are all in proper order, held by the power of love, so should the relations between countries, individuals, and spouses be directed.

These ideas may profitably be brought to an understanding of two works of medieval English literature. By the careful application of the Boethian ethic and its extensions through number and music symbolism

31. Trans. Richard Green (New York, Bobbs-Merrill, 1962,) p. 41.

to Chaucer's *Troilus and Criseyde* and the anonymous thirteenth century *Sir Orfeo* I feel some added understanding of these works can be achieved. In my discussion of these works, moreover, I will make references to other literature so that the theories discussed may be better understood as the intellectual context for a variety of writers. In almost all cases, there are direct connections with the text of Boethius.

The concept of the order of the universe plays a key role in the definition of Troilus' tragedy, a tragedy which consists basically in his failure to adjust his emotions properly. However much sympathy we may accord him as a tragic victim, his failure is ultimately a moral one. The underlying metaphor of the ethical framework within which Chaucer places Troilus is taken from arithmetic and musical discussions of proper order and these discussions are derived from the disciplines of the *quadrivium* which also pervade the *Consolation of Philosophy.*

The unhappiness of Troilus is intimately involved with love and its proper or improper attachments. Whatever he may strive after, his private behavior must conform to the basic law of order derived from proper proportion. The right proportion of musical intervals in its first and literal meaning, the Boethian *musica instrumentalis,* consists of proper and pleasing combinations of sounds made by musical instruments. When extended to the proportions of the human body, these pleasing combinations define, in terms of musical metaphor, the beauty of that human person in which physical elements or spiritual elements are ordered in relation to each other. This is the order of the *musica humana.* Chaucer's Black Knight in the *Book of the Duchess* implies these two orders of beauty in his description of Blanche:

> I knew on hir noon other lak
> That al hir lymmes nere pure sewynge
> In as fer as I had knowynge.[32]
>
> (lines 958-960)

Moreover, she is full of goodness (985), and truth has chosen her as its dwelling place (1005).

The Boethian *musica mundana*[33] is the final metaphor of proper proportion. *Musica mundana,* in its physical significance, indicates the proper

---

32. All Chaucer quotations are from *The Works of Geoffrey Chaucer,* ed. F.N. Robinson (Boston, Houghton Mifflin Co., 1957).
33. For a discussion of the three types of music, see David S. Chamberlain, »Philosophy of Music in the *Consolatio* of Boethius«, *Speculum,* 45 (1970), 80 - 97.

ordering of elements in the corporeal universe. We perceive its manifestation in the orderly sequence of the seasons, the peaceful coordination and balance of the elements, fire, air, earth, and water, and in the regular motions of planets and stars. According to Boethius, this order expresses a music not perceptible to the physical ear; it is comprehended by the intellect alone, and when hampered by the passions of the body, the intellect may not properly grasp it. For its perception, Boethius appeals to the mind's eye, the seat of intellectual perception, which is surrounded and by and fed by the bodily senses.

> There are various steps in and certain dimensions of progressing by which the mind is able to ascend, so that by means of the eye of the mind, which (as Plato says) is composed of many corporeal eyes and is of higher dignity than they, truth can be investigated and beheld. This eye, let me say, submerged and surrounded by the corporeal senses, is in turn illuminated by the disciplines of the *quadrivium (De Institutione Arithmetica,* Book I, chap. 1).

But as there is a spiritual dimension to man's proportions, so there is a spiritual side to the order of the universe, and this we see elaborated in the eighth poem of the second book from the *Consolation.* The contents of this poem emerge often in the literature of the Middle Ages. Peter Dronke[34] describes this passage and another like it *(Consolation,* Book III, poem 9) in terms of its relationship to Dante. Extensions of this Boethian idea may be seen in Shakespeare and Sidney; in fact, by the Renaissance it has become a commonplace. Dryden's »Hymn on St. Cecilia's Day« is a late and striking embodiment of how music is the ultimate metaphor of universal order.

The spiritual significance of order is epitomized in the proper directing of love. The Boethian moral framework regarding the terms of love, the necessity for love, and the unhappy consequences of ill-ordered love are obviously geared into the mechanism of tragedy in the *Troilus.* Chaucer's central statement of love's necessity and function occurs at the opening of Book III. It immediately sets up the perspective of the moral universe, the hierarchy of being, the relationship of the elements, and the necessary binding force of love in the entire framework:

34. »L'Amor che move il sole e l'altre stelle,« |in English| *Studi Medievali,* 3rd ser. pt. 1 6, (1965), 389 - 422.

In hevene and helle, in erthe and salte see
Is felt thi [Venus] myght, if that I wel descerne;
As man, brid, best, fissh, herbe, and grene tree
Thee fele in tymes with vapour eterne.
God loveth, and to love wol nought werne;
And in this world no lyves creature
Withouten love is worth, or may endure.
(lines 8 - 14)
Ye [Venus] holden regne and hous in unitee;
Ye sothfast cause of frendshipe ben also;
(lines 29 - 30)

This part of the poem represents only the culmination of the significance and effect of love in the world of the *Troilus.* As early as Book I, lines 253 - 59, Troilus refused to recognize his need for love properly ordered and thereby initiated a course of action that resulted in his undoing. We are reminded again of his dangerous indifference in Book 2, lines 1380 -90 and the importance of the law of love is repeatedly made clear:

That love is he that alle thing may bynde,
For may no man fordon the lawe of kynde.
(Book I, lines 237 - 38)

Troilus is bestial and fails to perceive proper order--the proportion and balance between mind and heart, intellect and passions. He fails as well to achieve a wholesome relationship with Criseyde, and the law of love and universal order is not apparent to his understanding. When Pandarus chides him for melancholy, he speaks (albeit unknowingly) of this higher ethical strucure:

Or artow lik an asse to the harpe,
That hereth sown whan men the strynges plye,
But in his mynde of that no melodie
May sinken hym to gladen, for that he
So dul ys of his bestialite?
(Book I, lines 731 - 35)

This bestial condition is redefined with reference to Troilus' mouse-like cowardice (Book III, line 736) and his senseless bull-like rage (Book IV, line 239).

The bestial dullness of Troilus is indicative of his progressively enfeebled moral and spiritual condition. Being out of tune and out of proportion makes him less of a man. He weeps, moans, and tosses helplessly in bed; he quakes and cries pitifully »Mercy, mercy, swete herte!« at the sight of Criseyde (Book III, line 98). Pandarus accuses him of cowardice and, taken aback at his repeated swooning, Criseyde asks:

> Is this a mannes game?
> What, Troilus, wol ye do thus for shame?
> (Book III, lines 1126 - 27)

Chaucer, like Boethius, appealed to the order of the universe for a proper understanding of the relationship between men and women. The fruits of such relationships should be happiness and concord. At the beginning of Book III, Chaucer's paraphrase from the Boethian picture of cosmic love is in ironic juxtaposition with the events of the Trojan War and the illusory happiness of Troilus with Criseyde. Troilus is unable to understand the discrepancy between his disordered love and the proper love exemplified in the order of nature as it is spelled out for us by Boethius (whose *Consolation of Philosophy* Chaucer had translated into Middle English). That his love for Criseyde should bring him to a tragic end neccesarily, as the inevitable outcome of the improper ordering of his love, is an important aspect of the poem.

Eventually, all things fall into a clear perspective for Troilus. After deteriorating to a point of extreme melancholy, suffering from delusions and hallucinations, he loses the desire to live. Slain in battle, Troilus gives up his soul which is thus finally freed from his body and its passions. He looks on the world from the eighth sphere to which he has ascended. With the model of proper order before him,

> ...he saugh, with ful avysement,
> The erratik sterres, herkenyng armonye
> With sownes ful of hevenyssh melodie.
> (Book V, lines 1811 - 13)

Then Troilus scorns his former misdirected passions. It was his initial misunderstanding for the necessity of a properly ordered love that caused his tragedy. In the palinode, Chaucer brings the world of his poem back to order, and the hero understands the causes of his undoing.

The appeal to order, the musical metaphor, the power of love and grace derived from a Boethian definition may be seen also in the thirteenth century anonymous poem, *Sir Orfeo*. Without discussing the specific problems of how this Breton Lay adopted the classical Orphic story and mingled with it the Celtic elements obvious in the setting of the Fairy King and his underworld,[35] one may focus on the activity of King Orfeo.

---

35. I have discussed these matters in »The Christian Music of *Sir Orfeo*,« *Classical Folia*, 20

His ability to perform on the harp with great skill operates metaphorically in the context of the poem. His musical ability is associated, by long tradition passed on in the Orphic legends and stories, with the spiritual life, the power of grace, and with the harmony of the spirit and the other world. The harp especially had long been associated with such power. It was the Pythagorean model of the harmoniously proportioned sound producer, and the vibrations of its strings corresponded to the sounds of the planets in their celestial spheres. The harp becomes the key to the personality of Orfeo insofar as he is the savior of his wife and effects the salvation of Heurodis through his spiritual power. It is a spiritual power which has consequences for his entire kingdom. By his skill in playing, the goodness of Orfeo is defined and symbolized.

The use of music as a definition of the hero's moral stature is commonly found in medieval literature. In the thirteenth century *Sir Tristrem* by »Thomas,« for example, it performs a function closely related to the use of music in *Sir Orfeo*. Music is given a prominent place in Tristrem's education (lines 287-304). His extraordinary skill at the instrument is acknowledged by all his rivals and no competition can stand up to him (lines 557-59). His musicianship culminates with the rescue episode (lines 1794-1925) which involves an Irish earl, Ysonde's admirer. This man comes to the court of King Mark in disguise, and by power of a musical instrument (as well as the cooperation of Mark's stupidity), wins the fair Ysonde from the court. Tristrem returns from his hunt and sets out in pursuit of the disguised minstrel. The power of his music draws Ysonde to him:

> His gleal for to here
> the leuedi was sett onland
> To play bi the riuere;
> therl ladde hir bi hand;
> Tristrem, trewe fere
> Mirie notes he fand
> Opon his rote of yuere,
> As thai were on the strand
> that stounde
> thurch that semly sand
> Ysonde was hole and sounde.

(1974), 3-20. Related material may be read in John Block Friedman, *Orpheus in the Middle Ages* (Cambridge, Mass., Harvard University Press, 1970). See also my review of Friedman in *Cithara* 10, No. 2 (1971), 105-108.

> Hole sche was and sounde
> thurch vertu of his gle.[36]
>
> (lines 1882-94)

Tristrem is the savior of his beloved, and music is the symbol for the power of his love and its redemptive efficacy.

There is an obvious resemblance between Sir Orfeo and Tristrem. After Orfeo leaves his kingdom, he undergoes a period of trial, desolation, and selfdenial. He takes with him the harp which becomes the image of his more than natural power over his environment. It is able to draw to him the wild beasts of the desert and order them tamely around the king. Orfeo uses the instrument, and his music, to charm animals and bring order to a realm filled with menace. By the music of his lyre, he establishes harmony with the hostile elements first in the desert and then in the lower world. His performance is so compelling that the king of the underworld offers to give him anything he asks. The power of Orfeo's lyre and his song restore Heurodis to the kingdom from which she was lost.

The role of music is not exhausted with Orfeo's return. By his harp he is known to the faithful steward:

> the steward loked, & can to se,
> & knewe the harpe wel blyve:
> 'Mynstrel!' he sayde, 'As thou most thryve,
> wher had thou that harpe, & howe?'[37]

The steward's confusion is resolved and the identity of the king who has been restored to his realm is established. The harp and its music were Orfeo's sole consolation in the desert, the powerful redemptive force against the world of fairy, and the index of the steward's faithfulness. Music, in this poem, becomes the means of drawing out the truth, of giving consolation, and of defining the hero's moral stature as a Christian.

Dante could hardly have chosen a better model of perfection on which to structure the levels of the celestial hierarchies than the spheres of the planets. From antiquity the skies were studied as the embodiment of unchanging and ideal concord, and in the ascent of the poet with Beatrice through the heavens there is a marked Boethian characteristic. A key passage occurs in the twenty-second canto where Dante has traversed seven

---

36. Here I am quoting from the edition of *Sir Tristrem* by George McNeill (Edinburgh, Blackwood and Sons, 1886). See also Madeleine P. Cosman, *The Education of the Hero in Arthurian Romance* (Chapel Hill, University of North Carolina Press, 1966), pp. 23-27.
37. *Sir Orfeo*, ed. A. J. Bliss (Oxford, Oxford University Press, 1966). Of the parallel texts, I quote from the Harley manuscript, lines 479-82.

spheres; from the eighth, he has a vision familar to us from a long medieval tradition, especially from the *Consolation*.[38] As in many such instances, the narrator is lifted to a prominent perch and acquires a philosophical
wisdom which enables him to voice the allegorical meaning of the voyage
he has just completed:

> My sight through each and all of the seven spheres
> Turned back; and seeing the globe there manifest,
> I smiled to see how sorry it appears;
>
> And I approve that judgment as the best
> Which least accounts it, and that man esteem
> Most worthy, who elsewhere brings his thoughts to
> rest.[39]

(Canto 22, lines 133-138)

The philosophical vision of detachment is linked in the tradition of Boethius with the ascent through the seven Liberal Arts, as the iconography of
the Liberal Arts instructs us. The vision afforded by philosophy after the
ascent up the ladder of seven rungs is a combination of Boethius' ideas
from the *De Institutione Arithmetica* and the *Consolation*. The equation
of the seven Liberal Arts with the ascent through the virtues and the
spheres does not lack iconographic counterparts. One example may be
seen in the bas-reliefs of the bell tower at the cathedral in Florence designed by Giotto in the mid-fourteenth century.[40]

This survey of number theory and the ethical context of literary
works hardly does justice to its full possibilities. It has been my intention
to demonstrate the permeation of number theory into literary contexts
and to emphasize the difference between this kind of study and that which
deals in mystical numbers. As a method of literary analysis it offers a wider intellectual context for the understanding of medieval works.

38. Cf.*Troilus*, Book 5, lines 1814-20; *Somnium Scipionis*, chapters 19-20. For more discussion and extensive bibliographic references, see John W. Conlee, »The Meaning of Troilus'
Ascension to the Eight Sphere,« *The Chaucer Review*, 7 (1972), 27-36.
39. Trans. L. Binyon, *The Portable Dante* (New York, Viking Press, 1947), p. 484.
40. Verdier, p. 324.

DE INSTITUTIONE ARITHMETICA: COMMENTARIES AND DERIVATIVE WORKS

This survey of commentaries and treatises on number theory in the Boethian manner is intended as an important adjunct to the analysis of the arts in earlier parts of this introduction. The primary purpose of the survey is to highlight the significance of Boethius' reputation among writers interested specifically in mathematical studies. It should demonstrate the direct impact of Boethius on late medieval and Renaissance number theory.

The body of material I approach here is large and relatively untouched by scholars. A more complete inventory of commentaries should eventually appear. Here I will list as many works as necessary to give on adequate notion of such treatises. This survey is restricted to printed works since, though seldom found in modern editions, they are certainly more accessible than manuscript commentaries. No doubt a large body of Boethian commentary remains to be studied in manuscripts. The material to be treated breaks down into the following broad categories: 1) summaries of the Boethian arithmetic significant in some respects for their treatment of number theory; 2)treatises on parts or all the system of the Liberal Arts, or some adaptation of it, which stress its developmental aspect; 3) discussions of proportions; 4) religious and allegorical applications of number which draw on Boethius.

Summaries of the *De Institutione Arithmetica,* though none covers its total scope, include works by Jordanus Nemorarius, Johannes de Muris, Luca Paciuolo, Faber Stapulensis, and Heinrich Glarean.

Jordanus Nemorarius, d. 1236, *De Arithmetica.* Published in Paris, 1496, with Boethius' *De Institutione Arithmetica* and the commentary of Faber Stapulensis.[41] Though there is no acknowledged relationship between Jordanus and Boethius, it is clear that both are concerned with the same principles of number theory. Jordanus could hardly have escaped knowing Boethius' work on arithmetic. The following outline is adapted from a schematic comparison between Jordanus and Boethius in the 1496 edition:

41. The most useful catalogue of these treatises is David Eugene Smith, *Rara Arithmetica* (Boston, Ginn and Company, 1908). This treatise is described on p. 65. I owe a debt of gratitude to the custodians of the rare book room at Columbia University where I was able to examine these books.

Topic	Boethius Chap.	Boethius Liber	Jordanus Propositio	Jordanus Liber
Even Number	3	1	1	7
	4	1	10	7
			12	7
Odd Number	4	1	3	7
			11	7
Even Times Odd	10	1	2	1
			3	1
			33-35	7
Odd Times Even	11	1	37-38	7
			40	7
Primary and Secondary Numbers	14	1	1-2	3
			27	7
Primary to Itself, Secondary to another	16	1	12	3
			15	3
Superparticular	24	1	37-38	9
			52	9
Superpartient	28	1	42	9
			53-54	9
Superparticular	29	1	43	9
The Pyramid	22	2	27-28	7
Heteromecic Number	26	2	17	7
			38	9
Arithmetic Proportion	42	2	2-3	1
			16	2
			1,5	10
Geometric Proportion	44	2	1,3,5,25,26	2
			38	9
			20	10
Harmonic Proportion	47	2	32	3
			34,36,40	10

Roger Bacon, c. 1220-1292, *Opus Maius* (ed. J.Bridges, Oxford, 1897). Bacon is widely recognized as a severe critic of the medieval educational system. In his *Opus Maius* he advocates a solid grounding in the study of numbers and although Boethius was already generally known, Bacon cites the *De Arithmetica* to emphasize the importance of mathematics. Roger Bacon is more interested in practical mathematics than in number theory, but he accepts the Boethian principles that all learning must begin with mathematics and that mathematics is essential to a true understanding of all disciplines, both natural and divine *(Opus Maius* IV, Dist. I, chap. 2).

Johannes de Muris, c. 1310-1360, *De Arithmetica*. Published in Vienna, by J. Singrenium 1515 (Smith, p. 117). Johannes de Muris was better known for his work in music. Until recently, he was erroneously consider-ed the author of the important *Speculum Musicae*. For theoretical discus-sions of both mathematics and music, de Muris acknowledges his depen-dence on Boethius. Though his treatise is only eighty-eight pages long, it contains considerable original material.

Luca Paciuolo, *Summa de Arithmetica*. Published in Venice, 1494, reprinted in Toscalano, 1523 (Smith, p. 54). The *Summa* with its Italian text and Latin chapter titles is a close paraphrase of sections from Boethi-us' *De Institutione Arithmetica*. It holds a significant place among these mathematical works since it is one of the few and certainly one of the ear-liest in the vernacular. Paciuolo's learning makes him a classical source for number theory, and he makes citations from Euclid in Greek. His even-tual interest in number is philosophical rather than practical, and he emph-atically states the necessity for the developmental study of mathematical disciplines, presumably through a scheme like the quadrivium. Drawing on the *De Institutione Arithmetica of Boethius, Book 1, chp. 1,* he states the purpose of mathematical study:»...*e tutti modi a solvere ogni caso pro-posto per algebra e de la proportione de proportionalita e partiree multipli-care summare sotrare per quelle a ogno bisogno perspectivo: musico, astro-logo, cosmographo, architecto, legisto, et medico.« (...it is in every way to solve each case proposed by algebra and to solve each proportion and pro-portionality and to divide, multiply, add, or subtract in every need related to music, astrology, cosmography, architecture, law, and medicine).*

Wolphganus Hopilius et Henricus Stephanus, Faber Stapulensis, *Epitome Boethii* (Paris, /1503) (Smith, pp. 80-81). There are many sides to the intellectual career of Faber Stapulensis, or Jacques Lefèvre d'Estaples. He was a reformer, a theologian and a mathe-matican. The combinations of the commentaries and studies of Faber Stapulensis and Jodochus Clichtoveus on Boethius and Jordanus resulted in a number of separate publications. The *Epitome* appeared alone in 1480. In 1496, it was published again with the *De Arithmetica* of Jordanus. In 1503, Jodochus Clichtoveus added his commentary to the *Epitome*. These works appeared in later printings combined with other treatises on practical arithmetic, geometry, and perspective.

Heinrich Glarean, *De Vi Arithmeticae Practicae Speciebus* (Paris, Jacobus Gazellus, 1543) (Smith, p. 1971). Like many other authors of treatises on practical mathematics, Glarean begins his discussion with a foundation in philosophical number theory and his exposition takes on a characteristically Boethian tone. After the definition of number and a division of numbers into categories, Glarean clarifies his definitions and categories with quotations from the *De Institutione Arithmetica*. He then proceeds to practical matters of addition and subtraction, not related to classical number theory. His treatment of proportion, ten pages long in this edition, borrows heavily from Boethius to whom he refers as his master, the »Divus Severinus« (p. 75).

Heinrich Glarean plays a significant role in the textual history of Boethian treatises. From his edition of the Boethian mathematics in 1546, a noteworthy edition for its time, Migne took the text for the *De Institutione Arithmetica* (vol 63, cols 1079-1168) and the *De Institutione Musica* (vol 63, cols 1168-1299) in the *Patrologia Latina*.

Those writers who treat of the theory of number in the context of the Liberal Arts or some modification of that system are numerous. They emphasize a slightly different aspect of the Boethian tradition and develop to a significant degree the principles I have been elaborating in connection with iconography, architecture, and music. These writers include Georgius Valla, Johannes Foeniseca, Joachim Fortius Ringelbergius, Hudalrich Regius, Claude Boissière, Robert Recorde, and Domenico Delfino.

Georgius Valla, *De Arithmetica* (Venice, Aldus Romanus, 1501). Smith is certainly lacking in perception when he remarks of Valla's book: »There is nothing noteworthy in this treatment« (p. 72). Smith also fails to note that the work consists of two volumes, as may be seen from the table of contents, and the Smith-Plimpton collection at Columbia University unfortunately contains only the first.

Valla's interests were very broad; he published a work on the astrolabe and produced an edition of Euclid. His ultimate purpose in mathematical and scientific studies was the evolution of a moral philosophy. The treatment of arithmetic in this work makes it clear in the first chapter that mathematics is eventually to result in a perception of a moral and religious order. This is a concept familiar from Boethius. Valla says: »*His sane instructi adiumento finem petimus, qui est ut bene vivamus, Deo rerum omnium opifici gratias agendo ipsum totis colendo praecordiis adorando metuendo ab omni malo abhorrendo.*« (Wholesomely instructed with these disciplines, we seek one end: that is that we live well, giving thanks to God the creator of all things, worshipping him with all our feelings, adoring and fearing him and abstaining from all evil.) His treatment of the sciences proceeds systematically through arithmetic, music, geometry, mechanics, optics, astrology, philology and goes on to moral philosophy. Since Valla was a physician, it is understandable that he should extend number theory to the proper order of the physical body and its health as well as to the moral health of the soul. Though his discussion of arithmetic as such is brief, it is noteworthy for its thorough definition of multitude and magnitude, categories which Boethius treats in summary fashion.

Johannes Foeniseca, *Quadratum Sapientiae* (Augsburg, Auguste Vindelicorum, 1515) (Smith, pp. 119-20). The arithmetic portion of this work is brief (two out of twenty pages), but the treatment of the whole is clearly inspired by the tradition of Boethius in that it traverses, in developmental fashion, all seven of the Liberal Arts. Foeniseca makes it clear that he draws his inspiration from Boethius, associating his name with each of the quadrivial disciplines.

Joachim Fortius Ringelbergius, *Opera* (Leiden, Gryphium Lugdinum, 1531) (Smith, pp. 165-67). This is a large compendium of the sciences, in the style of Valla and Reisch, and extends for 680 pages. The work begins with a short version of the Boethian *De Institutione Arithmetica*.

Hudalrich Regius, *Utriusque Arithmeticae Epitome* (Freiburg, Stephanus Gravius, 1550) (Smith, pp. 181-82). After a Boethian definition of number and a division into even and odd, this work contains a brief discourse on proportion. Regius feels that arithmetic is the first in the order of mathematical sciences which include optics, perspective, and mechanics. In this matter, he follows a tradition, formed largely in a parallel to the Boethian series of the Liberal Arts, which outlines the progress of the practical or mechanical arts.

Claude de Boissiere, *L'Art d'Arythmetique* (Paris, Annet Briere, 1554) (Smith, pp.260-62). Though Boissière does not traverse the entire series of the Liberal Arts, his ultimate aim is to comprehend a moral philosophy. His attitude is Boethian in a manner that draws from the *Consolation* as well as from the *quadrivium*. Though he eventually treats of practical mathematics, Boissière states the moral purpose for his study at the beginning. In this changing world of uncertainties, he explains, with threats from within and without, the study of science can provide a measure of certainty in our understanding of nature. In moments of difficulties, the pursuit of mathematical studies can provide security for the mind. Boissière emphasizes the consolatory aspects of philosophical study to which one ascends through the *quadrivium*. »La tante excellente, divin et admirable providence tellement a constitutué et disposé toute creature en ceste machine du monde, qu'il n'y a nation tant habitée et brutallé, laquelle cognoissant la quantité et grandeur des choses crées, ne soit attirée à contempler l'indicible et non iamais louée essence du createur.«

Robert Recorde, *The Grounde of Arts* (London, Reginald Wolfe, 1558) (Smith, pp. 213-14). This work is notable because it is written in English and enjoyed great popularity, going through some twenty-eight editions during the sixteenth and seventeenth centuries. It is Boethian less in its treatment of mathematics and more in that it understands the purpose of number theory as moral. As such, it makes the study of mathematics crucial to the system of education. Recorde rapidly surveys the seven Liberal Arts and claims strongly that the study of mathematics and its development through the disciplines he outlines should help to solve the moral problems of his country. Its preface is a close paraphrase of the well-known moral exhortation of Boethius.

Domenico Delfino, *Sommario di Tutte le Scienze* (Venice, Gabriel Giolito, 1565) (Smith, p. 275). The material of this work resembles that of the *Margarita Philosophica*. Delfino is one of the many who made such a compendium of all knowledge, a summary of the arts, sciences, philosophy, and theology of the Middle Ages. It is the particular mode of this treatment which differs from the *Margarita* though its matter is similar. Delfino's treatment of the Liberal Arts assumes modest proportions in the rather large book (pp. 21-55 of 416 pages) but it has an important function. It is presented in the manner of a journey, obviously adapted from Dante, and the persona who narrates these pages is ascending a mountain. The Arts are presented in a description modeled on the climbing of the mountain of the *Purgatorio*. As the *discipulus viator* goes up the mountain at whose summit he will discover philosophical wisdom, he encounters the allegorical figures of the seven Liberal Arts. Each encounter takes the time of one day, and Delfino continues his ascent in the company of a figure, different each day, who represent one of the arts and who imparts to him the knowledge of that art. The allegories appear, however, not in the Boethian order, but in a sequence familiar from Martianus Capella. The purpose of having music appear last, instead of second as in Boethius, is clear as the traveler approaches the summit in the company of the allegory of music. At the peak, the celestial music of the spheres comes to the traveler's ears, and the metaphor of the highest order, the order of the universe, asserts the eventual purpose of the ascent through the Liberal Arts. Delfino's inspiration is indisputably Boethius, whose name he invokes on the way, particularly in connection with arithmetic (p. 47).

There is a smaller group of writers and commentators concerned exclusively with the study of proportions, a topic which takes up a major part of the Boethius *De Institutione Arithmetica* (Book 1, chap. 22-31; Book 2, chap. 40-54). These writers include Thomas Bradwardine, Gaspar Lax, and Johannes Fernelius.

Thomas Bradwardine, c. 1290-1349, *De Proportionibus.*[42] This is an important work on mathematics and represents considerable original thought. Its tribute to Boethius is brief, and it goes on immediately to more advanced materials. However, the basis of its study of proportional principles is clearly determined by the Boethian *De Institutione Arithmetica*.

Gaspar Lax, *Proportiones* (Paris, Hermendus le Feure, 1515) (Smith, p. 121). This work is rather remotely related to Boethius. I include it here because it is a study of proportions, but it derives its ideas mainly from Greek sources. It begins with a series of thirty definitions drawn from Euclid. By the Renaissance, many theoretical mathematicians had access to the orginal Greek materials, and these sources superceded the treatise of Boethius in proportional discussions.

Johannes Fernelius, *De Proportionibus Libri Duo* (Paris, Simon Colinaeus, 1528) (Smith, pp. 157-59). This is a short work (24 folio pages) and makes few additions to the Bradwardine and Boethius discussions on which it substantially relies and from which it quotes regularly.

The final category of those who have inherited the Boethian tradition is somewhat more vague and serves as a catchall for the remainder of the material which in some way seems suitable for inclusion. This division includes the works of Gerardus Ruffus, Jodochus Clichtoveus, and Josephus Unicornus. In the works of these writers, two elements are consistently found: the dependence of their work directly on Boethius and a concern with mystical or Biblical numbers in conjunction with philosophical number. They combine their discussions of Boethius with scripture references and the writings of the Church fathers. This confusion of mystical and philosophical numbers does not occur in Boethius.[43]

---

42. Ed. and trans. H. Lamar Crosby (Madison, University of Wisconsin Press, 1955).
43. Though its discussion does not properly belong here, one may consult in connection with

Among all the Boethian commentators, the work of Gerardus Ruffus maintains a singular position for its handsome edition[44] of the treatise and for Gerardus' own careful, lengthy, chapter by chapter discussion of its contents. This edition with its commentary appeared in only one printing (Paris, Simon Colinaeus, 1521) (Smith, p. 31) and probably did not become popular since it made severe demands on the learning and patience of its readers. The nature of the discussion varies, and in it Ruffus demonstrates familiarity with Aristotle, Euclid, Plato, and other Greek writers as well as with Cicero and Latin authors. He quotes all of these, using Greek citations with freedom. In his comparison with ancient sources, he provides diagrams on numerous occasions to show relationships in schematic fashion between the Boethian theories and ancient philosophical number theory. Ruffus belongs in this fourth category because his commentary includes not only the ancient thinkers but has as well many references to Christian thought, and quotations are given from the fathers of the Church. In his commentaries, Ruffus »Christianizes« Boethius by mixing Biblical number speculation with number theory. Each commentary follows the chapter for which it was intended and often runs three to four times the length of the Boethian text. The typeset of the commentary is smaller than that of the original. The contents of the commentary generally follow this format: a paraphrase (with expansions of some details), comparsion with the ancients, religious applications.

Josephus Unicornus, *De Utilitate Mathematicorum* (Venice, Dominicus de Nicolinis, 1561) (Smith, pp. 298-300). This work combines a survey of the disciplines in the *quadrivium* with a discussion of mathematical principles covering perspective, optics, military sciences, and architecture. The opening chapter contains a Boethian assertion of the moral end for the study of mathematics: »*Vos igitur, qui ad veram solidamque doctrinam aspiratis adolescentes et qui nuper ex grammaticis, gymnasiis exessistis si doctrinae nobilitate insignes fore estis percupidi, statim post haec Euclidinae lectioni animam intendite*« (f. llv). (Therefore you who as young people aspire to a true and sound learning and who have just

mystical numbers and the Boethian theories a study by Heinz Klingeberg on the mystical Gospel commentaries of Otfrid (9th century) in light of the Boethian arithmetic. See his article »Zum Grundris der Ahd. Evangeliendichtung Otfrids,« *Zeitschrift für deutsches Altertum und deutsche Literatur*, Bd. IC, Heft 1 (1970), pp. 35-45.

44. In Ruffus' edition, we should note that chapter numbering differs from Friedlein, Migne, and most manuscripts; as much as one fifth of the text is omitted. For an extended discussion, see my article in the *Sixteenth Century Journal*, (1979), 23-41

emerged from the grammatical studies and the public schools, if you are desirous of being instructed in the nobility of learning, turn your mind immediately after these things to the reading of Euclid.) Unicornus then proceeds to combine the religious teachings of the Christian Church with the philosophical and moral teachings of the ancients. There is little doubt that the essential framework of his ideas is Boethian, and he appeals specifically to the *De Institutione Arithmetica* on f. 12v.

Jodochus Clichtoveus, *De Mystica Numerorum Significatione Opusculm* (Paris, Henricus Stephanus, 1513) (Smith, pp. 94-95). The existence of this work is a clear indication that Clichtoveus made the distinction between the mystical and philosophical functions of numbers. He discusses number theory in his commentary on Boethius and Faber Stapulensis. While this treatise mentions the Boethian mathematics (ff. 1-2v), its focus is mainly on the Bible and the writings of the Churchfathers. Smith's description of this treatise is to the point and, as well, illustrates his tone with material not in his interest: »Clichtoveus discusses, as is usual among such writers, the religious significance of one *(Quid unitas numerorum fons et origo designat)* and the numbers of the decade. He also mentions several larger numbers which were supposed to have some scriptural significance, not forgetting, of course, 666 'the number of the beast'.«

MANUSCRIPTS CONTAINING THE
DE INSTITUTIONE ARITHMETICA

The following list of manuscripts respresents all extant medieval copies of the Boethian *De Institutione Arithmetica* that have come to my attention. Any such list is necessarily incomplete, but it does at least provide evidence of the considerable popularity of the treatise. In searching for these manuscripts, I have depended on catalogues of manuscript collections and on correspondence from librarians. Professor Menso Folkerts has supplied information about a large number of the manuscripts. These are among the documents he examined for the edition of the Pseudo Boethian *De Geometria.* Professor Jean Schilling contributed about a dozen items. I am grateful to both of them for their help.

Dates and foliation are indicated when they were available. Since I was not able to examine each manuscript carefully, the data are usually based on catalogue information.

BAMBERG, Bibliothek zu Bamberg, Ms Msc. H J IV 11, S. X, ff. 8-108.
. . . . Ms Msc. H J IV 12, S. IX, ff. 1 - 139
. . . . Ms Msc. H J IV 13, S. XI, ff. 2-92
. . . . Ms Msc. H J IV 14, S. X, ff. 2-64.
BARCELONA, Archivo de la Corona de Aragon, Ripoll 168, S. IX.
BASEL,   Öffentliche Bibliothek der Universität, AN III. 18, S. XII.
. . . . AN III, 19, S. X.
BERLIN,   Universitäts Bibliothek, Ms lat. fol. 601, S. XIV.
. . . . Ms lat. quart. 528, S. XII.
. . . . Ms lat. quart. 578, S. IX (frag.)
BERN,   Bürgerbibliothek, Ms A 91, 14. S. XI. (frag.)
. . . . Ms F 219, S. XI (frag.).
. . . . Ms 299, S. X, ff. 1-29.
. . . . Ms 538, S. XI-XII, ff. 1-36.
BOLOGNIA, Biblioteca Comunale dell'Archiginnasio, Ms A 142, S. XV,
              ff. 1-49.
BRUXELLES, Bibliothèque Royale, ms 5444, S. XI.
. . . . Ms 18497.
BUDAPEST, Orszagos Szèchènyi Könvtàr, Cod. lat. 3, S. IX, ff. 7v-86.
CALIFORNIA, Library of Robert B. Honeyman, Gen. Sci. Ms 6, S.
              XIV-XV, ff. 1-64.

CAMBRAI, Bibliothèque de la Ville, Ms 928, S. X, ff. 1-55

CAMBRIGDE, Fitzwilliam Museum, Ms 295, S. XV.

. . . . Library of Peterhouse, Ms 248, S. XI-XII.

. . . . Pembroke College, Ms 269, S. XII, ff. 1v-94v.

. . . . Trinity College, Ms 940, S. XII, ff. 4-60.

. . . . University Library, Ms Ii III 12 (1776), S. XI, ff. 1-60.

CESENA, Bibliotheca Malatestiana, Plut. XXVI, 1, S. XV, ff. 1-57.

CHARTRES, Bibliothèque de la Ville, Ms 45, S. XI.

. . . . Ms 46.

. . . . Ms 498, S. XII, ff. 86-114 (destroyed in 1944; now to be
seen on microfilm at the Bibliothèque Nationale, Paris).

CHARLEVILLE, Bibliothèque de la Ville, Ms 184, S. XII.

CLUNY, Bibliothèque de la Ville, Ms 279.

DARMSTADT, Hessische Landes- und Hochschulbibilothek, Ms 2640,
S. XIII-XIV, ff. 212-42.

DRESDEN, Sächsische Landesbibliothek, C 80, S. XV, ff. 24-71.

. . . . C 99a, S. XV, ff. 17-26 (frag.).

. . . . C 349, S. XVI.

. . . . Dc 181, S. XII.

DUBLIN, Trinity College Library, Ms D. 4. 27.

. . . . Ms H. 12. 12, no. 7

. . . . Ms 441, S. XIV-XV.

. . . . Ms 1442, S. XIV (frag.)

DURHAM, The Cathedral Library, Ms C. IV. 7, S. XII, 50-66 (commentary).

EINSIEDELN, Stiftsbibliothek, Ms 358, S. X, ff. 16-32.

EL ESCORIAL, a-IV-3.

FIRENZE, (FLORENCE), Bibliotheca Medica Laurenziana, Plut.
29.20, S. IX, ff. 1-97.

. . . . Plut. 29.21, S. XIII, ff. 1-57

. . . . Plut. 29.22, S. XIII, ff. 1-55.

. . . . Plut. 29.23., S. XIII, ff. 1-46

. . . . Plut. 51.14, S. XI, ff. 87v-127v.

. . . . Plut. 89, sup. 81, S. XII, ff. 1-60.

. . . . Strozzi 25, S. XII, ff. 1-46.

. . . . Bibliotheca Nazionale Centrale, Ms II. IX. 5, S. XV, ff. 1-
101.

. . . . Bibliotheca Riccardiana, Ms 139, S. XII/XIV.

GRONINGEN, Universiteits-Bibliotheek, Ms. 103, S. XVI, ff. 216-221v
(frag.).

HILDESHEIM, Dombibliothek, Beverina 743, S. XII.

.... Domschatz, Bernards »Liber Mathematicalis«, S. X/XI.

KÖLN (COLOGNE), Dombibliothek, Ms 83, S. IX/X.

.... Ms 185, S. X.

.... Ms 186, S. IX.

KRAKÓW, Bibliotheka Jagiellónska, Ms 1954, pp. 226-33.

.... Ms 546, pp. 183-99.

.... Ms 1865, pp. 3-37.

.... Ms 1927, pp. 201-25.

LEIDEN, Bibliotheek der Rijksunivesiteit, B.P.L. 2391ª (frag.).

.... Vos S. o. 61.

LONDON, British Museum, Arundel 339, S. XIII, ff. 1-31.

.... Burney 275, S. XIV, ff. 336-59.

.... Harley 549, S. XII-XIII.

.... Harley 1737, S. XIII, ff. 1-33.

.... Harley 2510, S. XIII.

.... Harley 7656, S. XIV, ff. 1-7 (frag.).

.... Lansdowne 842, B, S. XV, ff. 1-50v.

.... Lambeth Palace Library, Ms 67, S. XII, ff. 1-62.

LUND, Universitetsbibliotheket, Ms 1, S. X.

LUXEMBURG, Ms 60.

MADRID, Biblioteca Nacional, Ms 9088 (Aa. 53), S. XI.

METZ, Bibliothéque Municipale, Ms 1250, S. XI.

MILANO, Biblioteca Ambrosiana, Ms C. 128, inf.

.... E 86 sup., S. XV.

.... N. 258. sup.

.... B. 36 sup.

.... T. 253 inf.

.... T. 79 sup.

.... Biblioteca Trivulziana, Ms 646, S. XII, ff. 1-47.

MONTE CASINO, Biblioteca dell'Abbazia, Ms 189, S. XI-XII.

MÜNCHEN (MUNICH), Bayerische Staatsbibliothek, CLM 3517, S. X

.... CLM 6285, S. XI (frag.).

.... CLM 6405, S. XI.

.... CLM 13021, S. XII, ff. 1-26v.

.... CLM 14401, S. XI.

.... CLM 14601, S. XII.

.... CLM 18208, S. XIII.

.... CLM 18478, S. XI.

.... CLM 18764, S. X, ff. 3-78.

. . . . CLM 23512, S. XII, ff. 1-59.

NAPOLI, Biblioteca Nazionale, Ms IV. G. 68, S. IX, ff. 108-78.

. . . . Ms V. A. 14, S. XV, ff. 1-44.

. . . . Ms V. A. 15, S. XIII, ff. 1-66.

. . . . Ms VIII. C. 19, S. XV, ff. 276^v-346.

NEW HAVEN, Yale University Library, Thomas E. Marston Collection, S. XIII, ff. 1-34.

NEW YORK,Columbia University Library, Plimpton Collection, Ms 165, S. XIII, ff. 1-28.

. . . . Ms 166, S. XIV, ff. 1-37.

OXFORD, Balliol College, Coxe Ms 306, S. XI-XII, ff. 5-43.

. . . . Bodleian Library, Ms 309, S. XI, ff. 149-64.

. . . . Digby 98, S. XII/XIV, ff. 86-107.

. . . . Land. Lat. 54, S. XIII.

. . . . Savile 20, S. XI-XII, ff. 1-64.

. . . . Selden Supra 25, S. XIII-XIV.

. . . . Corpus Christi College, Ms 118, S. XII/XIV, ff. 57-101^v.

. . . . Ms 224, S. XII-XIV, ff. 1-41.

. . . . Trinity College, Ms 17, S. XI, ff. 43-99.

. . . . Ms 47, S. XII, ff. 48-71.

PADOVA, Biblioteca Antoniana, Ms 515, S. XIII.

PARIS, Bibliothèque Mazarine, Ms 4319 (1), S. XII.

. . . . Bibliothèque Nationale, Ms lat. 817, ff. 1-3 (frag.).

. . . . Ms lat. 6401, S. XI, ff. 87-158.

. . . . Ms lat. 6639, S. IX, ff. 73-151.

. . . . Ms lat. 7039, S. XII, ff. 95-130.

. . . . Ms lat. 7181, S. IX, ff. 1-81.

. . . . Ms lat. 7182, S. X-XI, ff. 1-72^v.

. . . . Ms lat. 7183, S. IX, ff. 1-15^v.

. . . . Ms lat. 7184, S. X, ff. 1-91.

. . . . Ms lat. 7185, S. X, ff. 1-40^v.

. . . . Ms lat. 7186, S. XI.

. . . . Ms lat. 7187, S. XI, ff. 1-31^v.

. . . . Ms lat. 7188, S. XI, ff. 1-72^v.

. . . . Ms lat. 7189, S. XII, ff. 1-50.

. . . . Ms lat. 7215, S. XIII, ff. 146-66.

. . . . Ms lat. 7359, S. IX, ff. 2-82^v.

. . . . Ms lat. 7360, S. XII, ff. 1-56.

. . . . Ms lat. 10251, S, IX, ff. 3-69.

. . . . Ms lat. 11241, S. XI, ff. 7-71^v.

. . . . Ms lat.11242, S. XI, ff. 1-84^v.

. . . . Ms lat. 13009, S. IX, ff. 1-51.

. . . . Ms lat. 14064, S. IX, ff. 1-83.

. . . . Ms lat. 14065, S. XIII, ff. 7-46^v.

. . . . Ms lat. 16201, S. XIII, ff. 6-34.

. . . . Ms lat. 17858, S. X-XI, ff. 3-26v.

. . . . Nouv. Acq. lat. 1614, S. IX, ff. 1-76v.

. . . . Nouv. Acq. lat. 3044, S. XI, ff. 1-37v.

PARMA,   Biblioteca Palatina, Ms Parm. 718, X. XIII, ff. 1-25.

PISA,   Biblioteca Cateriniana del Seminario, Ms 27, S. XII, ff. 1-54.

PRAGUE,   Universitní knihovna, Ms 1717, S. IX, ff. 1-46.

PRINCETON, Princeton University Library, Kane 50.

RHEIMS,   Bibliothèque de la Ville, Ms 975, S. XI, ff. 1-29.

. . . . Ms 976, S. XI, ff. 1-36.

SANKT GALLEN, Stiftsbibliothek, Ms 820, S. XI, ff. 283-310.

. . . . Ms 248, S. IX, ff. 3-56.

. . . . Vadianische Bibliothek, Ms 296, S. XII.

SAUMUR,   Bibliothèque de la Ville, Ms 3, S. XIII, ff. 1-3 (frag.).

TORINO,   Biblioteca Nazionale Universitaria, Ms D-V-37, S. XII/XIV.

. . . . Ms F. IV. 1, S. XIV.

. . . . Ms F. IV. 1, S. VI (?) (frag.).

TOURS,   Bibliothéque Muncipale, Ms 802, S. X.

TRIER,   Stadtbibliothek, Ms 1102-51.

VATICAN,   Vat. lat. 3106.

. . . . Vat. lat. 6017.

. . . . Reg. 72, S. XII, ff. 1-56v.

. . . . Reg. 1070, S. XII, ff. 1-55.

. . . . Reg. 1075, S. XII, ff. 1-38.

. . . . Reg. 1267, S. XIV., ff. 57-88v.

. . . . Reg. 1789, S. XII-XIII, ff. 1-40.

. . . . Ott. 1551, S. XIII, ff. 132-62.

. . . . Ott. 2069, S. XIV, ff. 52-81v.

. . . . Pal. 1341. S. IX, ff. 6-60.

. . . . Urb. 1359, S. XV, ff. 1-71.

. . . . Borg. 210, S. XII-XIII, ff. 1-60v.

VENEZIA, Biblioteca Nazionale, Ms lat. 271 (= 1642), ff. 1-66v.

. . . . Ms lat. 332 (= 1647), S. XIII, ff. 4-37.

. . . . Ms lat. 333 (= 1648), S. XV, ff. 1-70.

. . . . Ms lat. 334 (= 1649), S. XV, ff. 1-60v.

VERCELLI, Biblioteca Capitolare, Ms 138, S. IX-X.

. . . . Cod. DCXCII (DCCII 749), S. XII, ff. 1-56.

VERDUN, Bibliothèque Municipale, Ms 24, S. X.

VEROLI, Biblioteca Giovardiana, Ms 5, S. XIII.

WIEN, (VIENNA), Österreichische Nationalbibliothek, Ms 50 (Salis. 373), S. X, ff. 1-46.

. . . . Ms 55 (Rec. 248), S. X, ff. 1-92.

. . . . Ms 83 (Univ. 926), S. XIV, ff. 1-38; ff. 39-65 (contains two copies).

. . . . Ms 177 (Salisb. 10. C), S. IX/XIV, ff. 118-65.

. . . . Ms 297 (Phil. 555), S. XII, ff. 1-59.

. . . . Ms 2269 (Rec. 429), S. XI, ff. 141-52.

. . . . Ms 2463 (Rec. 1320), S. XIII, ff. 1-42.

WÖLFENBUTTLE, Herzoglichen Bibliothek, Ms 1129, S. XIII-XIV, ff. 10-51.

. . . . Ms 3549, S. XV, ff. 3-52.

WROCLAW (BRESLAU), Bibliotheka Uniwersytecka (Rehdigeriana), Ms 54, S. X, ff. 1-85.

For an inventory of some twenty-five early printed editions of the *De Institutione Arithmetica,* extending from 1488 to 1570, see Smith *Rara Arithmetica,* p. 27.

This translation of the *De Institutione Arithmetica* is based on the Friedlein edition which, however, is occasionally in error. Corrections of that text will be found in the notes. While the proper readings can usually be inferred from the context, I have also consulted two manuscript versions which I compared with Friedlein throughout. These are Vatican Library, Pal. Lat. 1341 (S. IX) and Vatican Library, Reg. Lat. 1075 (S. XII).

A VIEW OF BOETHIUS' LIFE AND WORKS

The few details of the life of Boethius which come to us are exasperating in their incompleteness. What we do know of his education, writing, and political career, though, make it clear that he was a man of sensitivity and learning as well as of political skill and deep concern for matters of state. By reason of his singular influence on the history of medieval thought well into the sixteenth century, he has been termed a »seminal thinker« and a »maker of the medieval mind.« This is in spite of the fact that the originality and scope of his work can hardly match that of many who came in succeeding centuries.

Boethius was of aristocratic temperament and background; he was born about 480 into the distinguished Anicii family which gave Rome a number of renowned statesmen. At the death of his father and while still very young, Boethius was adopted into the family of Symmachus, a noble and learned man for whom he had high regard and who had a profound and beneficial influence on the younger man's intellectual life. Boethius shows us something of that important relationship in the preface to the *De Institutione Arithmetica*. As his *Doktorvater,* Symmachus not only advised and encouraged the young Boethius, but also made heavy use of the red pencil in those early writings where Greek sources were still being summarized and original ideas came forth haltingly and cautiously.

Part of Boethius' importance in western thought is due to historical accident. He read Greek and appreciated the significance of Aristotelian philosophy. Had he lived out his years, his plan to translate and comment on Aristotle's work would have contributed to a medieval philosophy of quite another character. As it is, the translation and commentary on Porphyry's *Introduction to the Categories of Aristotle* and the translation of the four works of Aristotle's *Organon* (with commentaries on two of them) provided the texts for the study of logic in the medieval schools until the Greek texts were reintroduced six centuries later. As my introduction demonstrates, Boethius' work was practically the only avenue by which Greek mathematical and musical theory reached the thinkers of the Middle Ages. It was the singular prerogative of Boethius to cast these works into the form by which they became diversified into many medieval versions and applications unknown to the Greeks. Most important, it was to a large extent through the Boethian version of pagan learning that the slow and arduous integration of Christianity and ancient thought began to take place.

The diverse aspects of Boethius' life are closely related to each other. He married Rusticiana, the daughter of Symmachus, so that his master also became his father-in-law. Politics was as much a part of family life as Greek thought. His father had been Roman consul in 487; Boethius became consul in 510 at the young age of thirty and, from that time, he was closely involved with the Roman senate and its interests. His two sons subsequently served in the office as well.

Theodoric, king of the Ostrogoths, the villain in Boethius' tragic fall from power, appears otherwise as a just and capable ruler. He became Roman governor in 493 and succeeded admirably in maintaining law and order in the kingdom. The Roman senate was allowed to continue under his rule, but its powers were severely curtailed, and only shreds of its former greatness remained. All the details of Boethius' unhappy clash with Theodoric--unhappy chiefly for the loss of what Boethius' writing had promised--are far from clear and have beein much discussed.[45] A combination of political and religious conflicts seems to emerge as the cause of their falling out. Boethius is now generally assumed to have been a Christian and author of four theological tractates for some time thought to be erroneously assigned to him. If Boethius is indeed the author of these works, then Theodoric's Arianism as well as his heavy-handed control of the Roman senate would have ill fitted Boethius' interests. In 523, he was imprisoned and disgraced; in the following year he was executed. His last work, *The Consolation of Philosophy,* comes from that unhappy year when, at the peak of his powers, political misfortune cut short a remarkable intellectual career.

45. For a survey and extensive bibliography, see William Bark, »Theodoric vs. Boethius: Vindication and Apology,« *American Historical Review,* 49 (1944), 410-26 and several essays in L. Obertello, ed., *Atti del Congresso Internazionale di Studi Boeziani* (Roma, 1981).

Boethius, to Symmachus, his Lord, the Patrician

In giving and receiving gifts, such courtesies are rightly esteemed, especially if it will be clearly understood among those who consider themselves important, that nothing is given with more generosity nor anything is received with greater benevolence.[1] Considering these things, I myself have brought to bear the not so sluggish weight of my resources and there is nothing characterized by greater preparation for the deed than these [resources] when thirst for possession has begun burning, and nothing too lowly of merit when the victor mind has subjected those things trod upon; but those things taken from the richness of Greek writings we have brought over to the treasury of the Roman language. So thus the reason of my work will be clear, even to me, if the things which I have laboriously chosen from the teaching of wisdom will be approved by the judgment of a very wise man. You see therefore that the result of such labor awaits only your examination. Nor will it go to the public ear unless it rests on the support of a learned opinion. Nothing amazing should be seen in that this work, which attends to the discoveries of wisdom, relies on the final judgment of another, not of the author. This work of reason is in fact put forth by the author's abilities since it is constrained to undergo the judgment of a prudent man. But for this little gift I do not set up the same defenses which hang over the other arts, nor in fact is any science free from all its parts, having need of nothing, and based only upon its own supports so that it might not lack the helps which come from the other arts.

Indeed, in shaping out of marble, there is one labor in cutting away the stone, and there is another notion in forming the image: the beauty of the completed work cannot attend on the hands of the same artist.[2] But

---

1. It is worthy of note that Boethius uses the »precious style« in his preface. Characteristically found in the works of late Latin writers, such as Martianus Capella, it is a highly ornate and artifical mode of writing with involved syntax and sentence structure. It contrasts sharply with the simple and more direct style found in the body of the treatise.
2. Boethius has in mind the other disciplines of the *quadrivium,* whose principles develop from the mathematical basis established in the *De Institutione Arithmetica.* Eventually he sees the possibilities of the application of this mathematical basis to the arts in terms which are beyond the merely mathematical, and therefore he considers his number theory as an expression of the philosophy of the beautiful. Since the *quadrivium* is intended to prepare its

the tablet of the picture has been trusted to the hands of various workers; the wax gathered together into a rustic scene, the red ochre of the color examined with the expertness of merchants, the linen elaborated by industrious weavers--these represent a multiplicity of artistic materials. Is not the same seen in the instruments of war? Here a spear thrower makes himself sharp in skill with arrows, there a powerful blacksmith groans at a black anvil, and yet another man trades coverings put together by his own labor on the orb of a bloodystained shield. From so many arts one art is made up. The completion of our labor traverses a long distance to a very simple end. You alone will put your hand on the final work in which it will not be necessary to be concerned about anything from the consensus of other judges. Although this cultivated judgment will be completed by many arts, yet it will be summed up by one examination.

It is therefore fitting that you test this work, to see how much labor was given to this study by us, taken from long periods of leisure, to test whether the quickness of an expert mind comprehends the flights of subtle matters, or whether the leanness of an undernourished language suffices to handle those things which are expressed unhindered in obscure propositions. In this matter why should the advantages of a foreign judge be sought when you as a most learned person in both languages are able to state in only a single pronouncement as much as those expert in the Greek speech dare judge of us.

I do not restrict myself slavishly to traditions of others, but with a well formed rule of translation, having wandered a bit freely, I set upon a different path, not the same footsteps. Those things which were discussed in a rather diffuse manner by Nicomachus concerning numbers, I have put together with moderate brevity; those things which demanded a greater care of understanding, but are gone through quickly, I clarified with a small additional explanation and I have even used formulae and diagrams for greater clarity of matters. The careful reader will easily recognize that this involved for us many nights of labor.

Once I finished writing about arithmetic, which is the first of the four mathematical disciplines, it seemed to me that only you were worthy of the gift--and I understood much more that this work had to be without error. Although for you the tendency for forgiveness is easy, at times a se-

readers for absolute philosophical studies, these studies are also understood and anticipated here. The notion of traversing a series of arts or skills to a desired end is thus a dominat idea of the prefatory presentation.

curity which is suspect has feared that easy friendliness. I considered that nothing should be offered to so great a reverence which seemed not to have been elaborated by inventiveness, perfected by zeal, and seen worthy in view of a long previous preoccupation. Therefore I will not hesitate to ask that in your kindness for me you cut off redundencies, fill up openings, correct errors, and take up with marvelous effectiveness of the mind what I have duly said. This thought impelled the otherwise sluggish delay in counseling. Things which will please you will return many fruits to me. I know indeed with how much more zeal we love our own goods rather than those of others. Rightly therefore as I would have given golden sheaves to Ceres or ripe branches to Bacchus, so I have given to you the beginnings of a new work. Receive then our gift with paternal benevolence. Thus you will consecrate the beginnings of my labors with your most learned opinion and as I consider it, their author will not be attributed a greater merit than you do who have approved it.

Here begin the Chapter Titles of the First Book

Here end the chapter titles of the first book.

BOETHIUS

De Institutione Arithmetica

Here begins the *First Book.*

1. Proemium: the division of mathematics.

Among all the men of ancient authority who, following the lead of Pythagoras, have flourished in the purer reasoning of the mind, it is clearly obvious that hardly anyone has been able to reach the highest perfection of the disciplines of philosophy unless the nobility of such wisdom was investigated by him in a certain four-part study, the *quadrivium,* which will hardly be hidden from those properly respectful of expertness.[3] For this is the wisdom of things which are, and the perception of truth gives to these things their unchanging character.

We say those things *are* which neither grow by stretching nor diminish by crushing, nor are changed by variations, but are always in their proper force and keep themselves secure by support of their own nature. Such things are: qualities, quantities, configurations, largeness, smallness, equalities, relations, acts, dispositions, places, times, and whatever is in any way found joined to bodies.[4] Now those things which by their nature are incorporeal, existing by reason of an immutable substance, when affected by the participation of a body and by contact with some variable thing, pass into a condition of inconstant changeableness. Such things (since as it was said, immutable substance and forces were delegated by nature) are truly and properly said to be. Wisdom gives name to a science

3. The first chapter of this book is taken from the *Arithmetica* of Nicomachus, Book 1, chap. 1-5. That work appears in English as *Introduction to Arithmetic* trans. M.L. D'ogge, intro. F. E. Robbins and L.C. Karpinski (New York, Macmillan, 1926). Nicomachus however, does not mention a four-fold way ; Boethius here coins the term so important in the medieval Liberal Arts curriculum. The earliest manuscripts consistently use the version *quadruvium,* and *quadrivium* is a later spelling.

4. Boethius speaks of the ideas of these categories, distinghishing them from particular occurrences. For him, as for the earlier Pythagoreans, they are distinguished from material, bodily things, subject to change and therefore not real.

in terms of these things, that is, things which properly exist, whatever their essences may be.[5]

There are two kinds of essence. One is continous, joined together in its parts and not distributed in separate parts, as a tree, a stone, and all the bodies of this world which are properly called magnitudes.[6] The other essence is of itself disjoined and determinded by its parts as though reduced to a single collective union, such as a flock, a populace, a chorus, a heap of things, things whose parts are terminated by their own extremities and are discrete from the extremity of some other. The proper name for these is a multitude. Again, some types of multitude exist by themselves, as a three, a four, a tetragon, or whatever number which, as it is, lacks nothing. Another kind does not exist of itself but refers to some other thing, as a duplex, a dimidium, a sesquialtar, a sesquitertial,[7] or whatever it may be which, unless it is related to anything, is not in itself able to exist. Of magnitudes, some are stable, lacking in motion, while others are always turned in mobile change and at no time are at rest.[8] Now of these types, arithmetic considers that multitude which exists of itself as an integral whole; the measures of musical modulation understand that multitude which exists in relation to some other; geometry offers the notion of stable magnitude; the skill of astronomical discipline explains the science of moveable magnitude. If a searcher is lacking knowledge of these four sciences, he is not

5. Philosophy etymologically means the love and study of such wisdom. The use of the word »philosopher« (one who pursues wisdom) was first ascribed to Pythagoras. See Diogenes Laertius, *De Vita et Moribus Philosophorum*, Book 1, chap.12.

6. On the distinction between multiude and magnitude, see Proclus, *In Primum Euclidis · Elementorum Comentarii*, ed. G. Friedlein (Leipzig, Teubner, 1873), p. 36 and Jcob Klein, *Greek Mathematical Thought and the Origin of Algebra*, trans. Eva Brann (Cambridge, Mass., M.I.T. Press, 1968), pp. 10-11.

7. Duplex, a relation of 1:2; sesquialtar, 2:3; sesquitertial, 3:4.

8. The division of mathematical sciences into four branches comes to Boethius from Nicomachus and may be outlined in this fashion:

I. Science of number (a) As such, absolutely καθ' ἑαυτό *Arithmetic* (b) Relatively πρὸς ἄλλο *Music*	II. Science of quantity (a) At rest ἠρεμοῦν *Geometry* (b) In motion σφαιρική *Astronomy*

A similar sivision is found in Proclus. Theon, however, includes music under arithmetic, maintaining they are identical sciences. There are instances in Boethius' treatise where this distinction does not hold, and problems arise between the limits for the first and second disciplines. Euclid and Domnius (5th century) avoid the distinction. For discussions, see Theon of Smyrna, *Exposition des Connaissances Mathématiques Utiles pour la Lecture de Platon*, ed. and trans. into French by J. Depuis (Paris, 1892; reprinted, Bruxelles, 1966), pp. 26-36.

able to find the true; without this kind of thought, nothing of truth is rightly known. This is the knowledge of those things which truly are; it is their full understanding and comprehension. He who spurns these, the paths of wisdom, does not rightly philosophize. Indeed, if philosophy is the love of wisdom, in spurning these, one has already shown contempt for philosophy.[9]

To this I think I should add that every force of a multitude, progressing from one point, moves on to limitless increases of growth. But a magnitude, beginning with a finite quantity, does not receive a new mode of being by division; its name includes the smallest sections of its body. This infinite and unlimited ability of nature in a multitude, philosophy spontaneously rejects. For nothing which is infinite is able to be assembled by a science or to be comprehended by the mind. But reason itself takes this matter of the infinite to itself; in these matters, reason is able to exercise the searching power of truth. It delegates the boundary of finite quality to the plurality of infinite multitude, and having rejected the aspect of interminable magnitude, it demands in a defined area a cognition of these things on its own behalf.

It stands to reason that whoever puts these matters aside has lost the whole teaching of philosophy. This, therefore, is the *quadrivium* by which we bring a superior mind from knowledge offered by the senses to the more certain things of the intellect. There are various steps and certain dimensions of progressing by which the mind is able to ascend so that by means of the eye of the mind, which (as Plato says)[10] is composed of many corporeal eyes and is of higher dignity than they, truth can be investigated and beheld. This eye, I say, submerged and surrounded by the corporal senses, is in turn illuminated by the disciplines of the *quadrivium*.

Which of these disciplines, then, is the first to be learned but that one

9. On the primary place of mathematical studies, see Iamblichus, *In Nicomachi Arithmeticam Introductio Liber*, ed. Hermengildus Pistelli (Leipzig, Teubner, 1894), p. 9; Theon, pp. 25-27.

10. Boethius summarizes Nichomachus' quotation from *Republic*, Sec. 527D). Nichomachus cites from memory or from a corrupt text since his version, below, differs from any other known version:

»You amuse me because you seem to fear that these are useless studies that I recommend; but that is very difficult, nay, impossible. For the eye of the soul, blinded and buried by other pursuits, is rekindled and aroused by these and these alone, and it is better that this be saved than the thousands of bodily eyes, for by it alone is the truth of the universe beheld.« D'ogge, pp. 186-187. Theon of Smyrna quotes the same passage, p. 7.

which holds the principal place and position of a mother to the rest?[11] This is artithmetic. It is prior to all not only because God the creator of the massive structure of the world considered this first discipline as the exemplar of his own thought and established all things in accord with it; or that through numbers of an assigned order all things exhibiting the logic of their maker found concord; but arithmetic is said to be first for this reason also, because whatever things are prior in nature, it is to these underlying elements that the posterior elements can be referred. Now if posterior things pass away, nothing concerning the status of the prior substance is disturbed--so, »animal« comes before »man«.[12] Now if you take away[13] »animal«, immediately also is the nature of »man« erased. If you take away »man«, »animal« does not disappear.

On the other hand, those things which are posterior infer prior things in themselves, and when these prior things are stated, they do not include in them anything of the posterior, as can be seen in that same term »man«. If you say »man«, you also say »animal«, because it is the same as man, since man is an animal. If you say »animal« you do not at the same time include the species of man, because »animal« is not the same as »man«.

The same thing is seen to occur in geometry and arithmetic. If you take away numbers, in what will consist the triangle, quadrangle, or whatever else is treated in geometry? All of those things are in the domain of number. If you were to remove the triangle and the quadrangle and all of geometry, still »three« and »four« and the terminology of the other numbers would not perish. Again, when I name some geometrical form, in that term the numbers are implicit. But when I say numbers, I have not implied any geometrical form.

The logical force of numbers is also prior to music, and this can especially be demonstrated because not only are numbers prior by their nature, since they consist of themselves and are thus prior to those things which must be referred to another in order to be, but also musical modulation itself is denoted by the names of numbers. The same relationship which we remarked in geometry can be found in music. The names diatessaron, dia-

---

11. See Plato, *Republic*, Sec. 522.
12. For the use of this argument, see Aristotle, *Topics*, Book 2, chap. 4.
13. Having shown to his satisfaction that number is metaphysically prior in creation, Boethius is now more concerned with logical priority. Hence here he means to »take away« or »abolish« in thought, logically. Nicomachus uses the Greek verb συναναιρεῖθαι with the same meaning, as does Aristotle in *Metaphysics* Sec. 1059B and Iamblichus, p. 10.
(τὰ δὲ συναναιροῦντα μὲν μὴ συναναιρούμενα δέ)

pente, and diapason[14] are derived from the names of antecedent numerical terms. The proportion of their sounds is found only in these particular relationships and not in other number relationships. For the sound which is in a diapason harmony, the same sound is produced in the ratio of a number doubled. The interval of a diatessaron is found in an epitrita comparison; they call that harmony diapente which is joined by a hemiola interval. An epogdous in numbers is a tone in music. I cannot undertake to explain every consequence of this idea, as to how arithmetic is prior, but the rest of this work will demonstrate it without any doubt.

Arithmetic also precedes spherical and astronomical science insofar as these two remaining studies follow the third [geometry] naturally. In astronomy, »circles«, »a sphere«, »a center«, »concentric circles«, »the median« and »the axis« exist, all of which are the concern of the disciplineof geometry. For this reason, I want to demonstrate the anterior logical force of geometry. This is the case because in all things, movement naturally comes after rest; the static comes first. Thus, geometry understands the doctrine of immovable things while astronomy comprehends the science of mobile things. In astronomy, the very movement of the stars is celebrated in harmonic intervals. From this it follows that the power of music logically precedes the courses of the stars; and there is no doubt that arithmetic precedes astronomy since it is prior to music, which comes before astronomy. All the courses of the stars and all astronomic reasoning are established exclusively by the nature of numbers. Thus we connect rising with falling, thus we keep watch on the slowness and speed of wandering stars, thus we recognize the eclipses and multiplicities of lunar variations. Since, as it is obvious, the force of arithmetic is prior, we may take up the beginning of our exposition.

2. Concerning the substance of number.

From the beginning, all things whatever which have been created may be seen by the nature of things to be formed by reason of numbers.[15]

---

14. A diatessaron is a relation of 4:3, an interval of a fourth, or an epitrita; a diapente is a 3:2 relation, an interval of a fifth, or a hemiola; a diapason is an interval of an octave; an epogdous is the relation of 9:8 or a whole tone interval. For a definition of these terms, see Boethius, *De Institutione Musica*, Book 1, chap. 16-19.

Boethius begins his discussion of logical priority in the arts with arithmetic and geometry because logical priority is more obvious in the case of those disciplines, not because geometry follows arithmetic. He makes it amply clear earlier that music must follow arithmetic.

15. See Nicomachus, Book 1, chap.6.

Number was the principal exemplar in the mind of the creator. From it was derived the multiplicity of the four elements, from it were derived the changes of the seasons, from it the movement of the stars and the turning of the heavens. Since things are thus and since the status of all things is founded on the binding together of numbers, it is necessary that number in its own substance maintain itself evenly at all times, permanently, and that it not be composed of diverse elements. What substance would one join with number when the model of it itself holds all things together? It seems to have been composed of itself alone. Nothing can be considered as composed of similar parts or composed of things which are joined without reasonable proportion. Numbers are discrete of themselves and differ from every other substance and nature. But it becomes evident that number is composed of parts, not similar parts, nor of those things which adhere to each other without reasonable proportion. There are, therefore, first principles which join numbers together, which are in accord with its substance and which are always permanent. Nothing can be made from that which does not exist, and things from which something is made must be dissimilar but must possess the capacity of being combined. These then are the principles of which number consists: even and odd. These elements are disparate and contrary by a certain divine power, yet they come forth from one source and are joined into one composition and harmony.

3. The definition and division of number; the definition of even and odd.

First we must define what number is.[16] A number is a collection of unities, or a big mass of quantity issuing from unities.[17] Its first division therefore is into even and odd. Even is that which is able to be divided into

16. Nicomachus, Book 1, chap. 7. For another definition of number which is derived from Boethius, see Jordanus Nemorarius, *Arithmetica,* Book 1, Preface; see also Martianus Capella, *De Nuptiis Philologiae et Mercurii,* ed. A. Dick, revised by Jean Préaux (Stuttgart, Teubner, 1969), Sec. 743-49; 768-71; Theon, pp. 29-31.

17. Boethius, following Nicomachus, gives a three-fold definition of number:

1) »Limited Multitude«. According to this definition number is merely a species of the genus multitude with the differentia limitation. For a similar definition, see Artistotle, *Metaphysics,* Book V, chap. 13. Eudoxus used the same definition, according to Iamblichus (p.10).

2) »A Combination of Monads«. This definition is found in Theon (p.29) and, according to Iamblichus (p.10), is as old as Thales.

3) »A Flow of Number, composed of Monads«. As such, number is a stream, and it flows from monad. This definition has been attributed to Moderatus of Gades a Pythagorean. See Stobaeus, *Eclogae Physicae,* Book 1, chap. 2 and Robbins and Karpinski, pp. 114-5.

equal parts without one coming between the two parts. Odd is that which is unable to be divided into equal parts unless the aforesaid one should come between the parts. This kind of definition is common and well known.[18]

4. The definition of even and odd number according to Pythagoras.

Yet the definition of number is different according to the teaching of Pythagoras. An even number is that which at the same and single division is able to be divided into very large parts with small spaces or into a very small number of parts, with large spaces, according to the contrary properites of these two types.[19] An odd number is one to which this cannot happen but whose separation into two uneven parts is natural. Here is an example. If some given even number is divided, neither part is found to be larger than the other; there is nothing but a separation into halves. No quantity is smaller when a division of each of these halves into equal parts is made. In this way the even number 8 is divided into 4 and 4. There cannot be any other division which would divide this term into a smaller number of parts since there are no fewer parts than two. Now when a whole is divided in a three-fold division, the total space within each part is diminished, but the number of the divisions is increased. What was said about the contrary natures of the two types of numbers is relevant to this kind of situation. We have previously explained that a quantity grows into infinite pluralities, and that a space, which may otherwise be called a magnitude, can be diminished into infinitely small sections and that this occurs in contrary ways. This is the case in the division of an even number: when the space is maximum, the quantity is minimum.

5. Another definition of even and odd, according to a more ancient method.

According to a more ancient method, there is another definition of an even number. An even number is that which can be divided into two equal or two unequal parts, but in neither division is there an even number mixed with an odd number or an odd mixed with an even number, except-

18. On Even Number, see Jordanus, Book 7, prop. 2, 10 and 12; on Odd Number, Book 7, prop.3, 10, 16.
19. Here is described the contrariety of magnitude and quantity. Halves are the greatest possible parts of a term in magnitude, yet there is a smaller number of them in a whole than of any other fractionl part; see Iamblichus, pp. 11-13. For the »Lamboid Diagram« which illustrates this principle, see D'ooge, p. 191.

ing alone the principal even binary number which cannot be divided into two unequal parts because it consists only of two unities. This number is the first equality, two. What I am saying is this: if one takes an even number, it can be divided into two even parts, as ten is divided into fives. It can also be divided into unequal parts, as when the same ten is divided into 3 and 7. In this manner, when one part of a division is even, the other part is even; and if one part is odd, the other part, which is also odd, when added to it does not make a total of more or less than ten. When ten is divided into fives, or into three and seven, and these portions are divided further, only odd numbers result. If, however, this number or any other even number is divided into even parts, as when eight is divided into 4 and 4, or in the same manner into odd numbers, as when the same eight is divided into 5 and 3, this is the result: in the first division both parts are even, and in the second both are odd. If one part of a division is even, the other cannot be odd, and if one part is odd, the other cannot be even. An odd number is that which is divided into other odd numbers by means of any division; numbers always show both types and neither type of numbers is ever able to exist without the other, but one must be understood as even, the other as odd. If you divide seven into three and four, one part is even, the other is odd. This same condition is found to exist in all odd numbers, and it can never be otherwise in the division of an odd number. These are the two types which naturally make up the power and substance of number.

6. The definition of even and odd in terms of each other.

Now if these types must be defined in terms of each other, the odd number may be said to be that which differs from the even by a unity, either by increase or reduction. In the same way, an even number is one which differs from an odd number by a unity, either by increase or reduction. If you subtract one from an even number or add one to it, you make it odd; if you do the same thing to an odd number, an even number is created thereupon.

7. Concerning the primary nature of unity.

According to the natural arrangement of things, every number is half of the sum of the numbers which come before and after it.[20] And if the

20. Nichomachus, Book 1, chap. 8; see also Jordanus, Book 1, prop.2. The theory may be stated algebraically as

numbers beyond these, whose midpoint is the given number, are added together, then the given number would be half of their total. This is true of the number above that joined to the one below the previous lower number down to the end which is unity. So, if one takes the number 5, the numbers around it are, above, 4, below, 6. If these are joined, they add up to 10 whose half is 5. The numbers which are beyond these, that is beyond 6 and 4, namely 3 and 7, if joined, also equal twice the number 5. Also the numbers beyond these, if they are joined together, are double five. These are two and eight. If they are joined, they add up to 10, whose half, again, is five. The same thing happens with all numbers, and the process can be repeated to the terminal point of unity. Only unity does not have next to it two terms, and so unity is half of that number which is next to it and of it exclusively. Next to one, only two is naturally placed, and unity is evenly a half of two. For that reason it constitutes the primary unit of all numbers which are in the natural order and is rightly recognized as the generator of the total extended plurality of numbers.

8. The division of even number.

There are three types of even number. One is called even times even, another even times odd, and a third odd times even.[21] The contrary types seem to be in opposite places, and these are the even times even and even

$$n = \frac{(n-1) + (n+1)}{2}$$

Anyone discussing the history of ancient mathematics should heed Jacob Klein's warning that the use of modern mathematical symbols distorts the fundamental concepts of early mathematics (p.5). I have used such symbols sparingly and only when I felt they would clarify an otherwise obscure text. D'ooge and Dupuis may be consulted for a fuller use of such symbols. It should also be noted that Boethius uses Roman numerals (for which I substitute Arabic) or the written word for a numeral, apparently without a consistent reason for taking one over the other. I have followed his designation in each instance.

21. This classification is found in Euclid, *Elements,* Book 7, where he gives four kinds of numbers: even times even, even times odd, odd times even, and odd times odd. Since the fourth type is not an even number, it is omitted by Nicomachus and Boethius. See Iamblichus, Book 1, chap. 87 and Theon, pp. 41-43. Jordanus gives this classification in Book 7, prop. 29, 31, and 32. See also Nesselmann, *Die Algebra der Griechen* (Berlin, G. Reimer, 1842), p. 192 and T. L. Heath, *A History of Greek Mathematics* (Oxford, Clarendon Press, 1921), vol. I, p. 70. Theon defines even times even as:

a) That number produced by the multiplication of two numbers.

b) That number all of whose parts are even, down to unity.

c) That number which has none of its designations in terms of odd numbers.

times odd. The middle type which participates in both the other two is the number called odd times even.

9. Concerning the number even times even and its properties.

The even times even number is that which is able to be divided into two equal parts, and those parts again into two equal parts, and then into further equal parts. This is done until the division of parts arrives naturally at indivisible unity. So the number 64 has a half of 32, and this has a half of 16, and this a half of 8; from here the four, which is a double of the binary, is divided into equals. The binary is divided by the half of unity and unity is naturally singular and does not accept division. We can see that however many parts of this kind a number may have, the term even times even applies to each of them and the same is true of their sum. So it thus seems to me that this number is called even times even because all of its parts, both in name and quantity, are found to be even times even. How and in what name and quantity this number would have even parts, we shall say later. Their generation, however, is thus: whatever numbers you write in double proportions after one, these are always generated as even times even numbers. It is impossible that they be born otherwise than through this process of generation. This matter may be shown by means of a descriptive example. Let there be put down all the doubles after one: 1, 2, 4, 8, 16, 32, 64, 128, 256, 512. If an infinite progression were made in this fashion you would find all the numbers in this progression. They are derived from one in double proportion and all of them are even times even numbers. Moreover, it is worthy of more than slight consideration that each of its parts is given the same name as all other parts of the series, namely even times even, and includes as great a total perfection of the quantity as is the number part of the quantity of even times even which it contains. It is such that the parts of a number correspond to each other so that however great one part is, the other has the same quantity, and however much that part is, it follows that it is necessary that the sum of that multitude be discovered within the former number.

If the arrangement of the numbers is even, the number series functions so that the two middle numbers of a series correspond to each other, and those two which are beyond the first two correspond to each other, and this kind of distribution can be continued until each end of the series reaches its terminus. Therefore the order of even times even from one to 128 may be written out as follows: 1, 2, 4, 8, 16, 32, 64, 128 with 128 as the highest number. In this kind of distribution therefore, because the distribution is even, no middle number can be found. There are two middle numbers, that is 8 and 16, which must be taken into consideration insofar as they correspond to each other; if you multiply one by the other their product equals the highest number in the series, 128. The eighth part of 128 is 16, and the sixteenth part is 8. Beyond these two numbers are two others which correspond to each other in the same way; these numbers are 32 and 4. Now 32 four times equals the highest number, 128, and so does 4 times 32. If we go beyond these numbers, we see that 64 is half of 128, and two is one sixty-fourth of that same number. This is true also of all the other numbers in the series up to the extremes; certainly all these numbers enjoy the same correspondence. Thus the product of the first and last number is 128.

If, however, we take an uneven set of numbers according to the nature of uneven sets, one median term can be found of the same (I designate the »sum« or the »end« in the same way) and the median number corresponds to itself. Let us set up this order in the following manner: 1, 2, 4, 8, 16, 32, 64. There is one median, the number 8. Eight is an eighth part of the whole sum, and it corresponds to itself in denomination and quantity. In the same manner, the numbers on both sides of 8 are given names which correspond to their particular quantities and exchange their terms. Four times sixteen is equal to the highest term in the series, and so is sixteen times four. Thus beyond these, 32 twice is equal to the highest term and so is 2 times 32, and all are equal to the term 64, just as unity times 64 is equal to 64. This therefore is what has been said before, that all parts of this order can be found to be even times even both in respect to their names and in respect to their quantities.

This basic ordering of numbers has come about through careful consideration and through the great constancy of divinity, so that when disposed in an orderly fashion, the minor sums in this number series are always equal to the number directly above them minus one. So if you add one to the number which follows, that is, to two, they make 3, that is, one less than four. And if to the preceding you add four, they add up to seven, which sum is overcome in the following eight by only a unity. But if you add that same 8 to those already accumulated, they make 15, which is in quantity equal to the number 16, except that a simple unity impedes it. The primary progression of numbers protects and maintains this order. Unity, which is first, is closed off from the following number two by unity alone. There is no reason to marvel at the fact that the total growth of the number series is in harmony with its principal nature to the highest degree. This consideration is especially related to the process of understanding those numbers which we will show to be perfect, imperfect, and superabundant.[22] The quantity of all the parts, collected together, may also be compared by means of the appropriate rules to the last number in the series.

We cannot absent-mindedly pass by the fact that in this number series, the largest extremity and final term of the same number series is also brought about by the parts within the series which correspond to each other when they are multiplied together. First, if the dispositions are equal, the middle numbers should be multiplied and then those beyond them and so on all the way out to the two ends of the series. If the distributions are even, according to the nature of even numbers, there will be two terms in the middle, as for example, in the disposition of numbers where the end term is 128. In such a disposition, the medians are 8 and 16; if they are multiplied together, they increase to the largest number in the series. Eight times 16, or sixteen times 8: if you multiply these, you get 128. The numbers which are beyond them in that series do the same thing; four and 32, if multiplied together, yield the same sum. 32 times 4 or four times 32, by an unchangeable necessity, reach 128. This will happen all the way to the extreme terms, that is 1 and 128. One times the last term equals 128; one hundred and twenty-eight multiplied by unity is none other than the same term.

If the distribution were made of an odd number of terms, there would be only one middle term, and this would be multiplied by itself. In an order of numbers where the final term is 64, there will be only one middle term, that is 8; if you multiply eight by itself, it will give 64, and those terms beyond the middle term will give the same total, and so with those beyond that point. Four times 16 is 64, and 16 times four gives the same result. Again, two times 32 arrives at a figure no less than 64, and 32 times two is the same. One times sixty-four gives the same number, without variation.

10. Concerning the even times odd number and its properties.

The even times odd number is one which has itself taken on the nature and substance of an even number, yet it is put in a class contrary to the nature of the even times even number.[23] It will be shown by means of a different kind of calculation how far it may be divided. When a number is even, it can be divided into even parts, and these parts will not emerge as indivisible and irreducible, as are 6, 10, 14, 18, 22 and other numbers like

22. See below, Book 1, chap. 19.
23. Nicomachus, Book 1, chap 9; Jordanus, Book 1, prop. 2 and 3; Book 7, prop. 33-35.

these. But if these numbers were further divided into two parts, you would then come to odd numbers, which you would not be able to divide into equal parts. It happens that these numbers have all their parts designated in a manner different from that designating the quantities of those same parts. It can never be that any part of such a number takes the designation or quantity of an even number at the same time. If its division is even, its quantity would be odd; if its division is odd, its quantity would be even. The half or median part, which is an even division, of the number 18 is 9, which is an odd quantity; a third of this number, 18, which is an odd division, is 6, which is an even amount. Once more, if you convert that number into six parts, which is an even divison, each of these parts will be three, and three is an odd term. The ninth part, which is an odd term, is two, an even number. This same is found in all the other numbers which are even times odd. It can never happen that the name and the number of its part be the name and number of its category.

This number series may be generated if numbers are disposed from one so that each number in the series is two more than the one before it, yet they lie in the natural sequence and order of all odd number series. Now if these are multiplied by a binary number, the product will bring about an even times odd number in a proper manner. Let us posit then the first unity, one. After this comes that number which differs from the first one by two, that is, three. After this, put that which is two more than the previous one, namely 5, and so on *ad infinitum,* so that there is this kind of dispositon: 1, 3, 5, 7, 9, 11, 13, 15, 17, 19. These numbers naturally following each other are odd, and among them appears no even number. If you multiply these by a binary number, you would do it in this way: twice 1 is two; this total can be divided one time, but its parts are found to be indivisible because of the nature of indivisible unity. Then multiply twice 3, twice 5, twice 7, twice 9, twice 11, and from this process you will generate the following numbers: 2, 6, 10, 14, 18, 22. If you divide these, they will take one division and refuse any other because a second division into halves is excluded by any even times odd number.

In these numbers there is only a distance of four between each, as between 2 and 6 there is the number 4; the same four makes the difference between 6 and 10, between 10 and 14, and between 14 and 18. Each of these surpasses the other by the number four, and that happens because in the first order of numbers put down, each number preceded the following by two. Those numbers were then multiplied by two and they created the progression of four-fold numbers, since 2 multiplied by 2 yields the sum of four. Therefore according to the disposition of the natural number, each even times odd number is in the fifth place after the one of its kind which comes directly before it;[24] four numbers precede each number in the series if we include the even times odd number before it, and three numbers exist between the two. They have been created by means of a number that contains 2 after unequal numbers have been multiplied. These are said to be contrary species of numbers then--even times even and even times odd-- because in the even times odd number the greater extremity only can take division; in the even times even only the smallest term is free from being divided.

The product of the even times even numbers beginning with the outside terms and going to the middle which is contained within these extremities is the same as the product of the terms just within the extremities, and this product is the same as the product of the two median numbers and the products of corresponding numbers progressing from the middle in even distributions. If there were an odd disposition assembled with one median number, this number would be created in relation to parts placed on either side, and the series would progress from there to the extremities. For example, in the disposition of numbers which are even, such as 2, 4, 8, 16, the same total is created when 2 is multiplied by 16 as when 4 is multiplied by 8; in either case, they make 32. Now if there is an odd number of terms in a series, such as 2, 4, 8, the extremes multiplied make the same total as the middle multiplied by itself. In an even times odd number, if there is one term in the middle and terms are placed around it, then if they are added together in appropriate pairs, they will yield the same product as that given by the middle term added to itself. It is the same with those which are beyond these terms, of which these are the middle, and so on to the extremities of the series so that in this order of even times odd, 2, 6, 10, the 2 joined with the 10 gives twelve, whose half is found to be 6. This happens if the two middle numbers are added together and are equally distant from the end terms set up beyond them, as in the series, 2, 6, 10, 14. Two added to 14 grows to 16; six joined to ten gives the same number. This figuring in numerical terms at the middle is consistent until it comes to the extremes.

24. These numbers are distant from each other by five if one counts the beginning and terminal numbers. The Boethian desire for order here seeks out hidden orders within orders. (See note 31 below.)

11. Concerning the odd times even number and its properties; its relation to the even times even and even times odd.

The odd times even number is put together from each of the two previous and is inclosed in a middle place by the two extremities so that as a result, by means of that which is lacking from each of the other two, the knowledge of odd times even may be assembled.[25] It is the kind of number that can be divided into equal parts and even parts of these parts can be divided, but that separation into equals cannot be continued all the way to unity, as it can be with numbers such as 24 and 28. These can be divided into halves and their parts again reduced into other halves, without a doubt. There are some other numbers whose parts can take still further division, but this divison never comes to unity. Therefore insofar as it is capable of more than one division, the odd times even has a similarity to the even times even yet differs from the even times odd. But insofar as the division is not carried to unity, it does not show dissimilarity from the even times odd and is distinct from the even times even.

It also happens that this number has two characteristics which those discussed before do not have, and it shares two with them. It has what they do not have since in the case of even times odd only the highest term is divided while in the case of even times even only the smaller term is not divided; in the case of odd times even not only is the largest term capable of division, but the smaller term is not exempt from it. Further, its parts are divided, and yet they cannot be divided down to unity, but before unity a point is reached at which you are not able to divide. It is also true that odd times even numbers are able to take divisions which certain of its parts respond to and that these are called even or odd according to their generic types which are derived from their quantities, according to their similarity to an even times even number; other parts take a name which is different from their own quantity, that is, according to the fashion of even times odd. Now in the number 24 the even quantity of a fraction is named by an even number. For in the fourth part, 6, the second part, 12, the sixth part, 4, and the twelfth part, 2, the quantities and the part are of the same kind. Yet some numbers are opposed, as the third part, 8, and the

---

25. Theon of Smyrna defines the odd times even number (pp. 41-43) as one produced by the multiplication of an odd by an even number, which has even halves when divided by 2 but on further division yields odd parts. See Nicomachus, Book 1, chap. 10; Euclid, *Elements*, Book 7, def. 9; Jordanus, Book 7, prop. 37, 38, 40. For a discussion, see Heath, *Euclid's Elements*, vol. 2, pp. 281-84.

eighth part, 3. In the twenty-fourth part, one, when the name of the part is even [twenty-four] the quantity is found to be odd [one]; and when the quantity [24] is even, the name is odd.

These numbers are generated in such a manner that their substance and nature are designated by their generation, and they are born out of even times even and even times odd. Just as all even times odd numbers are generated from series of odd numbers in natural order, so even times even numbers are generated from a progression of numbers successively doubled. If odd numbers are disposed in natural order from three and under these is placed a series of numbers beginning with four and successively doubled in this fashion, this will result:

3	5	7	9	11	13
4	8	16	32	64	128

Now with these numbers arranged in this manner, if we multiply the first number in the top row by the number directly below it, we will have a number containing three fours; again the first number times the second below it gives us a number containing three eights; or again the first multiplied by the third is a number containing three sixteens; and so on until the end. If you take the second number in the top row and multiply it with the first below [5 x 4], or if the second is multiplied by the second [5 x 8], or the second by the third [5 x 16], and the same mulitiplication is carried on to the end, or if you take the third and multiply it by the first [7 x 4] and go to the end and do the same with the fourth and with all the numbers in the upper order, then all the numbers created will be odd times even.

From such a situation as this then let us take an example: if 3 is multiplied by four, they make 12; if five is multiplied by four, the number 20 is produced; and if 7 is multiplied 4 times, 28 grows out of it, and so on until the end. Again if 8 is multiplied by 3, 24 is born of it; if eight is multiplied by five, their product is 40. If 8 is multiplied by 7, their product is 56, and if in this manner all the lower numbers are multiplied by the upper, or the upper multiplied by the lower, then all those numbers which are born from these unions will be found odd times even.

Such is the admirable scheme of the odd times even number that when there is a disposition and clear description of numbers, along the latitude are found the properties of the even times odd number and along the longitude the properties of the even times even. In the latitude of the given diagram, the sum of the two middle numbers is equal to the sum of the extemes. Or in the case of one middle number, when doubled, it is equal to the sum of the extremes. In the longitude, in truth, the diagram describes the properties and nature of the even times even number. The product of the two mean numbers is equal to that derived from the multiplication of the two extremes, or the product from the mean squared is equal to that which is expressed by the multiplication of the extremes. Now this diagram which is placed here is made in this manner (p 88). Whatever numbers in the order of even times even have been multiplied by the number 3, and whatever numbers come from it, are put in the first row. Again those which are born of the multiplication by five are in the second row. After that come the numbers which are created from the multiplication of the even times even numbers by seven and these we put in the third row, and we have carried this same idea out in the remaining parts of the diagram.

3	5	7	9
4	8	16	32

12. An explanation of the diagram relating to the nature of the odd times even.

The reasoning of the diagrammatic description above is this: if you look at the latitude where there is a middle point between two terms, you can join these terms, and you will find them to be double the median, as 36 and 20 make 56, half of which is 28, which is the term placed between them. Again, if you join 28 and 12, they make 40, half of which is 20, and 20 is found to be their middle term. But where there are two middle terms, the two extreme terms joined together are equal to the middle terms, as 12 and 36; when you join them together, they become 48. If you apply their middle terms to each other, that is 20 and 28, the sum is the same. Also in another part of the latitude in the same order where numbers have been written down, in no way will the reasoning process differ. You will also note the same thing about the other numbers. This is done according to the diagram of the even times odd number, concerning the nature of which an explanation was given above.

Again if you look at the longitude, where two terms have one median, the product of the extremes multiplied gives the same total as the middle term multiplied by itself, for 12 times 48 gives 576. If the middle term, that is 24, is squared, it produces the same number, 576. Again if 24 is multiplied by 96, it makes 2304, and their middle term, that is 48, if multiplied by itself, produces the same 2304. Where, however, two extremities include two middle terms, the number produced by the multiplication of the outer extremities is the same as that produced by the multiplication of the middle terms. So, twelve multiplied by 96 makes 1152. The two middle

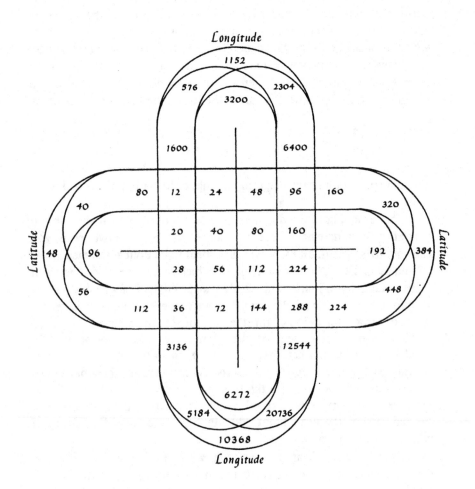

terms, that is 24 and 48, if multiplied, give the same 1152. This is done in imitation of and in agreement with the even times even number, so the participation of this inborn property is recognized in the odd times even. The same reasoning process and description may be noted on the side of the longitude. For this reason it is obvious that this number [odd times even] is produced from the two prior [even times even and even times odd] because it retains their properties.

13. Concerning the odd number and its division.

The odd number[26] is distinguished by nature and substance from the even number since the even is able to be divided into two equal halves; the odd cannot be so divided unless a unity comes between the halves, yet in a similar way it has three subdivisions.[27] Of these, one part is a number called primary and incomposite, another is called secondary and incomposite, and a third is connected at the midpoint to these other two and by nature draws from the understanding of the other two and is of itself secondary and composed, but when compared to the others, it is found to be primary and incomposite.

14. Concerning the prime and incomposite number.

The prime and incomposite number is that which has no other factor but that one which is a denominator for the total quantity of that number[28] so its fraction is nothing other than unity, and such are 3, 5, 7, 11, 13, 17, 19, 23, 29, 31. In these numbers, there is only one factor found, and each is a denominator of itself and that only once, as said above. In three is only one factor, that is, a third, which has a denominator of three, and unity is a third part of the number. In the same way, there is only a fifth part of five, and that is unity; this is found to be the case sequentially with each individual number. Such a number is called primary and incomposite because no number of this kind can be measured except through that factor which is the mother of all numbers, unity. A three cannot give

---

26. Nicomachus, Book 1, chap. 11.
27. A consistent definition and division of primary and secondary number is not to be found among the ancients. For Euclid, a prime number is one measured by unity alone (Book 7, def. 11) and so this includes the number two, which Nicomachus omits. Aristotle also includes the dyad *(Topics,* Book 8, chap. 2, sec. 57. a39). In Nicomachus and Boethius, primary numbers are always odd. See also Theon, pp. 37-39.
28. On prime and secondary numbers see also Martianus Capella, Sec. 750-752 and 772-775 and Jordanus, Book 3, prop. 1; Book 7, prop. 25.

name to 2, for if you compare just 2 alone to 3, it is smaller than 3, but if you double it, it becomes greater than 3, since it grows to 4. One number is a measure of another number as often as, either alone or doubled or multiplied by three, or however often that number is compared to another, its sum, neither smaller nor larger, comes exactly to the amount of the number it is compared with. So if you compare 2 to 6, the binary number is the measure of the larger number by three. Therefore, no other number can measure prime and incomposite numbers except unity alone because it is put together from no other number, but is produced only from unities, increased and multiplied by them. Three times one is three, five times one is five, and seven times one is seven, and so all the others which we described above are generated in the same manner. These multiplied by themselves as prime numbers create other numbers, and you will find them to be the first substance and power of things; the origin of all numbers is generated from them as though they were elemental material because they are uncomposed and formed by simple generation. Into them all numbers are resolved since all numbers are drawn from them; these numbers are not produced from others nor can they be reduced to others.[29]

15. Concerning the secondary and composite number.

The secondary and composite number [30] is itself an odd number because it was formed by the same properties as an odd number, but it retains in itself no elemental substantial principle; it is composed of other numbers and has parts named both in relation to itself and in relation to other terms. But you will find unity alone among these as the part named by itself and among those named by another name you will find evidently either one or however many others and however great the numbers will be. The secondary and composite number is produced by those which make it up, as are 9, 15, 21, 25, 27, 33, 39. Each of these has a factor which is a denominator of itself, which is unity, as in the case of 9, whose

29. Gerardus Ruffus, in his amply annotated edition of the Boethian *De Institutione Arithmetica*, comments on this chapter that such reasoning can be interpreted as an allegory of how man knows God. Numbers may be imperfect in respect to themselves but perfect in relation to others so the knowledge of man can be imperfect in itself but comprehend something of the perfect. »So our manner of knowing goes from composed things to simple, from the imperfect creatures to perfect beings, from created things to God« (f. 39).
30. Nicomachus, Book 1, chap. 12. Euclid defines a composite number as one measured by some other numer (Book 7, def. 13, 14). See also Theon, pp. 39-41 and Martianus Capella, Sec. 776-777.

denominator is one, and in the case of fifteen which is fifteen times one, and the same process occurs in the cases of all other listed above. But they can also be factored by another term, as 9 has a third, which is three, and 15 has a third, which is five, and a fifth, which is three; 21 has its third, which is seven, and a seventh, which is three, and the same order follows throughout the series.

This number is called secondary because it is not measured by unity alone but by another number as well, namely by that number with which it is made up, nor does it have in itself anything of an elemental substance, for it is produced from other numbers. Nine is produced from threes; fifteen from three and 5; 21 from 3 and 7, and the others in a similar manner. A number is said to be composed of another number when it is able to be resolved into that number of which it is said to be composed, namely into numbers which measure the composed number. Nothing which is able to be so broken down is uncomposed; such a number is said to be composed by the very necessity of things.

16. Concerning that number which in itself is secondary and composite but when related to another is primary and incomposite.

Now with these two numbers put next to each other, that is the primary and incomposite on one side and the secondary and composite on the other side, and separated by a natural difference, there is another number one should consider, and this is in the middle and is itself composite and secondary and again is able to take measurement, and therefore is capable of being called by a different name.[31] Even if it is compared to another number of a similar nature, it is not joined with that number by a common measure; nor does it have equal fractional parts, as do both 9 and 25. No common measure of numbers measures these terms, except perhaps unity, which is the common measure of all terms. Nor do these numbers have equal parts, for what is a third in 9, does not exist in 25, and what is a fifth in 25, does not exist in 9. Therefore these numbers are by

31. See Nicomachus, Book 1, chap. 13 and Jordanus, Book 3, prop. 12, 25. Both Euclid (Book 7, def. 13) and Theon (pp. 41-43) include this classification, but differ in their descriptions of it. Theon uses the terms »absolutely« and »relatively« prime. Boethius, following his source, establishes here the same kind of relationship we saw in the even times even and even times odd (Book 1, chap. 8). The third type is defined in terms of and partakes of characteristics from each of the previous two. See also Book 1, chap. 19 where Boethius defines perfect, deficient, and superabundant numbers. In Book 1, chap. 20 all three ranks are brought into play, as by a numerial architecture, in the production of the perfect number.

nature both secondary and composite but when compared to each other they become primary and incomposite because no other measure will fit both, except unity, which is a denominator for both; in nine unity occurs nine times, in 25, it occurs 25 times.

17. Concerning the production of primary and incomposite numbers, the secondary and composite number, and the number which in relation to itself is secondary and composite and in relation to another is primary and incomposite.

The generation and origin of these numbers occurs in the kind of investigation which Eratosthenes,[32] in particular, called a »sieve« because with all the odd numbers put in the middle, through the sieve by means of that art which we are about to pass on, each number is sifted out from the others which are seen to be of the first, second, or third types. Let all the odd numbers be disposed in an ordered series of any long sequence from the number three: 3, 5, 7, 9, 11, 13, 15, 17, 19, 21, 23, 25, 27, 29, 31, 33, 35, 37, 39, 41, 43, 45, 47. Now with the numbers distributed in this manner, we should consider what the first number is which can measure the first number in this order. Then it next measures the one placed two numbers ahead of the first and in order to measure the one after that, it must pass on two more counts, and so on in the same manner. If two more counts are made, the number that is measured once more is measured by the prime number. In the same manner, each number that is measured has two in between it and the previous number measured and this progression goes on from the first number to infinity.

But let me not do this in a general or confused manner. The first number measures that which has been placed two numbers after itself, and measures it by its own quantity. So three, with two numbers intervening, that is, 5 and 7, measures nine, and measures it by its own quantity, that is, three times, for three times the number 3 is nine. Then after nine, skip 2 numbers; go to the one which comes after it which should be measured by the first odd number in terms of the quantity of the second odd number in the series, that is by 5. So if after 9 we skip 2 numbers in the series, that is,

32. Eratosthenes was born in Africa during the third century B. C. He was called by Ptolemy to Alexandria to head a library there. His most important scientific and mathematical achievements included establishing a chronology of dates from the Fall of Troy and measuring the circumference of the earth. See M. Cantor, *Vorlesung über Geschichte der Mathematik* (Leipzig, Teubner, 1892-98), chap. 1.

11 and 13, the third number, 15, is measured in terms of the quantity of the second number in the series, that is, by the quantity of five, and three measures 15 five times. And accordingly, if beginning from 15, I go past the two others placed after it in the series, the first number measures it by the quantity of the third odd number in the series. If after 15, I skip 17 and 19, I come to 21, which is measured by the number 3 seven times. Of the number 21, three is a seventh part. And in doing this *ad infinitum,* I find that the first number of this series, if two others of the series are dropped, is able to measure each of the succeeding numbers in the order and by the quantity of an odd number in the series.

If in the case of the number 5, which is situated in the second place in the series, someone would want to find the first and following measures, four odd numbers should be skipped after 5, that is 7, 9, 11 and 13. After these comes the number 15, which the number 5 measures according to the quantity of the first odd number, that is, by three, for 5 measures 15 by three. Then if one skips four numbers after that, the second number of the series, that is 5, is measured by its own quantity. So after 15, with the numbers 17, 19, 21 and 23 put aside, we find 25, which the number 5 measures by its own quantity, since when multiplied by five, it grows to 25. After this, one skips another four numbers thus preserving the constancy of the same order, and the number which follows, the third of the series, that is the number 7, measures that quantity by five. And this process may be carried out in an infinite progression.

If the third number which is able to be measured is sought out and six places are left in between, the order comes to the seventh number; this is able to be measured by the quantity of the first number, that is, three. After that number, with six others put between, the number that the series then gives is able to be measured by five, the second in the series, and this measures 15 three times. If one should then skip another six numbers in between, that number which follows, the seventh [21] is able to be measured by the quantity of three in the factor of seven, and this established order proceeds to the final number of the series.

Therefore the odd numbers submit to the patterns of measurement according as they have been placed in an order conformable to nature. Thus are they measured: if beginning with even numbers we skip over the odd numbers, which are two apart at fixed intervals, then first there is 2, second 4, third 6, fourth 8, fifth 10. If any number doubles its place, then by means of that double place I can determine the number of terms skipped over in the series to reach that very number; so it is in the case of the number three, which is the first number in the series of odd numbers.

Since every first place is one, this place multiplied twice makes twice one, which is 2. So two numbers will be passed over to reach three. Again if the second number in the series, which is 5, doubles its position in the order [2 x 2], it will give 4, so it is necessary to pass over just that many numbers to reach 5. So also with seven, which is the third number in the series; if it doubles its position, it creates six, since twice three gives six. If the fourth number in the series of odd numbers doubles its place, the result is 8; the fourth number, 9, falls right after the eighth number. This process must be found in the remaining numbers of the series as well.

The series itself will give us the means of making measurements according to the order of arranged numbers. Whatever the first number measures, it measures according to the first number, that is, it measures according to itself. Then the first number measures the second by the second, and the third by the third and the fourth by the fourth. When the second number takes a measure, the first one which it can measure, it does so according to the first number, that is, by the second number in the series of odd numbers; it measures the third by the third, and in the remaining numbers this measure will persist with this same constancy. Therefore if you look at other numbers or others which are measured, or which themselves are measured by others, you will find there cannot be a common measure of all these, nor can all numbers exist in the same way. Some from this series are able to be measured by another number so that they are measured by one only; then others may be measured by many numbers. Yet there are some for which there is no measure but unity. We may consider those which are able to take no measure besides unity to be prime and incomposite numbers. Those numbers which have some measure more than unity or the term of a fractional part, those we may term secondary and composite. A third type is of itself secondary and composite, but when compared to another, it is primary and incomposite. One can obtain this type by the following reasoning.[33] If you multiply any of those same numbers by their own quantity, the numbers which result compared to each other will be joined by no common bond in terms of quantity. For if you square both 3 and 5, three squared will equal 9, and five fives will equal 25. There is no common bond for these in terms of measurement. If you examine the numbers which 5 and 7 create, you will

33. This method may also be seen in Euclid (Book 7, def. 1; book 10, def. 2) and is often referred to as the Euclidian way of discovering the largest common denominator.

see as well that these also will be incommensurate with each other. For five as we said, when squared, gives 25, and seven sevens make 49, and between these two there is no common measure, except that mother and creator of all numbers, unity.

18. Concerning the discovery of those numbers which in relation to themselves are secondary and composite but in relation to others are primary and incomposite.

We are able to find such numbers by the following reasoning;[34] if some one proposes such numbers to us and asks us to perceive whether they are measurable by a certain measure or whether only unity measures each of them with certainty, then the art of finding them is as follows: with two given odd numbers, take the smaller from the larger, and if the one which is left is larger, take the smaller one from it. If it is smaller, then take it from the larger one. This process must be continued until the final point of unity stops its path of reduction or until some other number [emerges] , odd by necessity, if odd numbers are proposed to each number. Then you will see that the numbers which remain are equal to each other. If this subtracting process comes to he number one, these numbers by necessity are called primary in relation to each other and are said to be joined by no other measure but unity itself. If, however, the process does bring us to some other number as said above,and incurs the end of the diminution, it will be that number which measures both amounts and that which remains is therefore said to be their common measure.

Let us say that we have two numbers given to us and that we are asked to discern whether they can be divided by any common measure. Let us take 9 and 29, for example, and make a reciprocal subtraction in the following way: let us take the smaller from the larger, that is nine from 29, which leaves 20. From this twenty, let us subtract the smaller number 9, and we will have 11. From this if we take 9, 2 will be left. If we take 2 from 9, there are seven left; and if we take 2 from that, there are 5, and another 2 from that gives three, which, if diminished by an addititonal 2, would leave only unity. Again if from two numbers we take away one, the terminus of the reduction will reside in the number one; of these two numbers, that is 9 and 29, there cannot be any other measure than the above. These two numbers we then call prime and incomposite in relation to each other.

---

34. See Jordanus, Book 3, prop. 12 and 15.

Let there be other numbers proposed for us under the same condition, that is 21 and 9, so that they may be compared to each other and what they are may be investigated. Again I will take the quantity of the smaller number from that of the larger, that is 9 from 21, and 12 will be left. If I subtract 9 fom this number, 3 is left. If another three is taken away from nine, six is left. If one takes three from that, 3 is left from which no three can be taken, and this three is equal to itself. Now the three which was subtracted reduces the original term to three, and since it is equal to the number of the other term, it cannot be diminished or taken from it. We therefore can say that these two numbers [21 and 9] are commensurable and that the number three which remains is their common measure.

19. Another division of even numbers according to perfect, imperfect, and superabundant.

As much an introduction to odd numbers as brevity permits is thus finished. Of even numbers there is a second division.[35] Some are superfluous, others diminutive, according to their types of inequality. Now every type of inequality may be considered to be figured either in terms of larger numbers or in terms of smaller numbers. The larger numbers surpass by means of an immoderate plenitude, in terms of the numerosity of parts, the size of their own total bodies; the smaller numbers, as though needy and oppressed by proverty, suffer a certain slight lacking of their nature. The sum of their parts makes them such as they are. Those larger numbers, whose parts extend themselves more than is enough, are called superfluous numbers as are 12 and 24. These, when compared to the sum of their parts factored out of the total body are found to be larger than that sum. Half of 12 is 6, a third part is 4, a fourth part is 3, and a sixth part is 2, a twelfth part is 1. The total sum $[6+4+3+2+1]$ amounts to 16. This surpasses the total of the entire body. Again half of the number 24 is 12, a third is 8, a fourth is 6, a sixth is 4, an eighth is 3, a twelfth is 2, and a twenty-fourth is 1. All of these numbers add up to 36. In this matter, it is obvious that the sum of the parts is greater than and exceeds the size of the original number. And so this number whose parts added together exceed the sum of the same number is called superfluous.

That number is called diminished[36] whose parts, when put together

35. Nicomachus, Book 1, chap. 14; Theon, pp. 35-37; Martianus Capella, Sec. 753.
36. Nicomachus, Book 1, chap. 16

in the same way, are exceeded by the multitude of the whole term, as 8 and 14. For 8 has a half, which is 4, and a fourth, which is 2, and an eighth, which is 1, all of which added together give 7, and this number is still confined within the total body of 8. Again 14 has a half which is seven, a seventh, which is two, and a fourteenth, which is one, the sum of which reaches the number 10. This is altogether smaller than the original term. So these numbers, those whose parts added together exceed the total, are seen to be similar to someone who is born with many hands more than nature usually gives, as is the case with a giant who has a hundred hands, or three bodies joined together, such as the triple-formed Geryon. Or this number is like some monstrosity of nature which suddenly appears with a multiplicity of limbs. The other number, whose parts when totaled are less than the size of the entire number, is like one born with some limb missing, or with an eye missing, like the ugliness of the Cyclops' face. Or the number is like one who is born naturally deficient in relation to some member, who emerges short of his total fulness.[37]

Between these two kinds of number, as if between two elements unequal and intemperate, is put a number which holds the middle place between the extremes like one who seeks virtue. That number is called perfect and it does not extend in a superfluous progression nor is it reduced in a contracted reduction, but it maintains the place of the middle; the sum of its parts is not more than the total nor does it suffer from a lack in comparison with the total, as are 6 and 28. Six has a half which is three, and a third which is two and a sixth which is one, and these numbers, if brought together to form a total sum, are found to be equal to the original number.[38] Indeed, the number 28 has a half of 14 and a seventh part of 4, and is not lacking in a fourth part, 7, and a fourteenth part, 2, and a twenty-eight part, one, all of which numbers brought together into one total are equal in parts to the size of the original term. Joined together, all these parts equal 28.

---

37. See Homer's description of Scylla, *Odyssey*, Book 12, lines 226-250. Geryon, whom Nicomachus does not mention, is the three-headed giant that ruled the island of Erytheia and was slain by Hercules in the tenth of his twelve labors. See *Theogony*, lines 287 ff. and F. Brommer, *Herakles: Die Zwölf Taten des Helden in Antiker Kunst und Literatur* (Munster, Manner, 1953), pp. 39-42. For excessive or or deficient numbers, see L. E. Dickson, *History of the Theory of Number* (Washington, Carnegie Institution of Washington, 1919), Vol. 1, chap. 1.
38. Euclid *(Elements,* Book 7, def. 22) defines the perfect number, τέλειος, as one equal to its own parts. Theon, p. 75, gives a similar definition but presents only 6 and 28 as examples.

20. Concerning the generation of the perfect number.

There is in these a great similarity to the virtues and vices. You find the perfect numbers rarely, you may enumerate them more easily, and they are produced in a very regular order.[39] But you find superfluous or diminished numbers to be many and infinte and not disposed in any order, but arranged randomly and illogically, not generated from a certain point. Within the first ten numbers there is only one perfect number, 6; within the first hundred, there is 28; within a thousand, 496; within ten thousand, 8,128. These perfect numbers always end in one of two numbers, 6 or 8, and these numbers always provide the final term in alternating fashion for the perfect numbers. First there is six, then 28, after this 496, which ends in 6 and which is the first number, then 128 which ends in 8 and that is the second number.

The generation and production of these numbers is fixed and firm and they cannot be achieved in any other way nor, if they are brought about in this way, is it at all valid to create them by another mode. Let the even times even numbers be disposed from one, and in order as far as you wish.[40] Then you will add together the second with the first, and if a pri-

---

39. Nicomachus mistakenly concludes that the final digit in perfect numbers alternates between 6 and 8. They do so up to the fifth order, but they are irregular after that. Heath gives a longer list of perfect numbers *(History,* Vol. 1, p. 75) which because of their length must be expressed algebraically:

I	$2(2^2 - 1) = 6$	
II	$2^2 2^3 - 1) = 28$	
III	$2^4(2^5 - 1) = 496$	
IV	$2^6(2^7 - 1) = 8,128$	
V	$2^{12}(2^{13} - 1) = 33, 550, 336$	
VI	$2^{16}(2^{17} - 1)$	
VII	$2^{18}2^{19} - 1)$	
VIII	$2^{30}(2^{31} - 1)$	
IX	$2^{60}(2^{61} - 1)$	[37 digits]
X	$2^{88}(2^{89} - 1)$	
XI	$2^{126}(2^{127} - 1)$	[63 digits]

Among ancient as well as medieval thinkers, virtue was commonly considered as a median between extremes. Aristotle uses mathematical examples to define the virtuous middle, then describes it:

»Virtue, therefore, is a mean state in the sense that it is able to hit the mean. Again, error is multiform (for evil is a form of the unlimited, as in the old Pythagorean imagery, and good of the limited) whereas success is possible in one way only (which is why it is easy to fail and difficult to succeed--easy to miss the taget and difficult to hit it); so this is another reason why excess and deficiency are a mark of vice, and observance of the mean a mark of virtue.« *(Ethics,* Book 2, chap. 6, sec. 13-14; trans. H. Rackham, Loeb Classical Editions, London, 1934).

40. For other accounts of this method, see Euclid, Book 9, def. 36 and Theon, pp. 75-77. For commentary and further references, see Heath, *Elements,* vol. 2, pp. 242-246.

mary and incomposite number is made from that addition, you will then multiply the sum by the second number you have added on. If from that addition, a primary number does not emerge, but a secondary and composite number, skip this number and add to it the next one which follows. If then you do not yet emerge with a primary and incomposite number, add another number and see what emerges. But if you find a primary and incomposite number, then multiply it by the number added on from the last sum. Now the even times even numbers may be disposed in this manner: 1, 2, 4, 8, 16, 32, 64, 128. Then work in this fashion. Put down one and add two, and see what number is made from this addition; from it comes 3, which of course is primary and incomposite. After unity, you add the number two. If you then multiply three, which is the number achieved by this addition, by two, which is the last number added on to the sum, then without a doubt a perfect number is born. Twice 3 makes six, which has one part which is its factor, namely six, and 3 is its half, while 2, the second number added on, is its third and these two [2 + 3] have been multiplied to give the product 6.

Twenty-eight is produced in the same way. If to one and two, which make three, you add the following even times even number, that is four, you arrive at seven. Then take the last number four, which you added in sequence, multiply the total by it, and a perfect number is produced. Seven times 4 is 28, which is a number equal to the sum of its parts; it has one as its factor, which is its twenty-eighth, and a half of 14, a fourth of 7, a seventh of 4, and a fourteenth of 2 and that corresponds to the medial term.

After these numbers are found, if you want to go on to discover the others, it is necessary that you pursue them by the same reasoning. It is necessary that you put down one, and after this 2 and 4, which add up to seven. The perfect number 28 showed itself a while ago by means of this process. The even times even which follows this number is 8; this is added on, increasing the previous number to 15. But this is not a primary and incomposite number, for it has another factor beyond the number by which it is named in itself, that is beyond the factor of one into fifteen. Because this number is secondary and incomposite, pass by it and add to the previous number the next even times even number, that is 16, which when added to 15, makes thirty-one. This is primary and incomposite. Multiply this number by the last number added on to the total, so that 16 times 31 yields 496. Now this sum is the perfect number within the order of 1000, and it is equal to the sum of its parts.

Unity is first in power and possibility but not in act and operation, and is itself perfect. If I take this first number from the proposed order of numbers I see that it is primary and incomposite because if I multiply it by itself, that same unity is produced for me. Also, the number one brings about only unity, which is equal only in potency to its parts, and is also like other perfect numbers in act and operation. Rightly therefore is unity perfect in its own strength, because it is prime and incomposite and when by means of itself it has been augmented, it maintains itself through its own strength.

We have now said enough about that quantity as it is in itself and we will move on to that part of our work which deals with quantity in relation to another.

21. Concerning a quantity that is related to another.

Whenever one quantity is related to another, there is a basic twofold way in which this may occur.[41] Any given thing in comparison to another is either equal or unequal with it. Something is equal when that which is compared does not fall short by virtue of its smaller sum nor exceed by virtue of a larger sum, as ten compared to then, or three compared to three, or a cubit compared to a cubit, or a foot compared to a foot, and so on with other measurements similar to these. When one part of a quantity is related to another quantity and there is an equality of parts, such a relationship is by nature undivided. No one can say that this is of such an equality and that equality is of another equality, for every equality serves one measure in its own proper arrangement. Any given thing which has a quantity compared to it is not known except by the other term to which it is compared. Just as a friend is a friend to a friend, or a neighbor is a neighbor to a neighbor, so is equal said to be equal to equal.

There is a double division of unequal quantity. It is divided since it can be unequal as larger or as smaller, and these relationships operate in denominations which are contrary to each other. The larger is greater than the smaller and the smaller is lesser than the larger and the two cannot be known by the same term, as we said about equality. But they are marked by diverse and separate natures accordingly as they are compared in contrary terms, as are a teacher and his student, a striker and the one struck, or whatever other contrary thing may be related to something in terms of opposites.

41. Nicomachus, Book 1, chap. 17. Relative quantity belongs more properly to a study of music, but the distinction is problematic in Nicomachus. See note 8 above.

22. Concerning the types of major and minor quantity.

Of major inequalities, there are five kinds. One is called multiplex, another superparticular, a third superpartient, a fourth multiplex superparticular, and a fifth multiplex superpartient. Now to these five types of major inequality are contrasted another five types of minor inequality. Each major type is opposed in parallel fashion to a minor type, and these types of minor inequalities are individually related to each of the majors types. The major were given above, so the minor are called by the same names, but distinguished by the prefix »sub.« They are called submultiplex, subsuperparticular, subsuperpartient, submultiplex superparticular, and submultiplex superpartient.[42]

23. Concerning the multiplex number, its types, and their generation.

The multiplex is the first kind of major inequality,[43] prior by nature to all the others and more distinguished than they, as it will be shown a little later. The multiplex is of such a type that, when compared with another, in contrast to the one it is compared to, it contains that other number more than once. This happens to the first number in the disposition of natural numbers. All those numbers which follow the number one maintain the sequence and variety of all the multiplex proportions. To the number one, that is to unity, two is a duplex, three is a triplex, four is a quadruple, and so on, until all the multiplex quantities are covered, proceeding in that order. That number is said to be more than one if it takes its beginning from the binary number and so proceeds on to infinity through the third, fourth, and on in the order and sequence of all the numbers. Contrasted with this type of proportion is distinguished that which is called the submultiplex and this is the first type of the minor inequaltities. This term is of such a type that when brought into comparsion with another, it numbers the sum of the larger, and beginning with its own quantity, equally is a factor for that term and proceeds equally. It is measured, I say, by the same way that it factors. If a smaller number measures a larger number twice, it is called subduplex, if it measures it three times, it is called subtriplex, if four times, subquadruple, and this process goes on *ad infini-*

---

42. These classifications are further defined in Thomas Taylor's *Theoretic Arithmetic* (London, A.J. Valpy, 1816). Theon (p. 75) gives two classifications, one of which coincides with that of Nicomachus. See D'ooge, p. 213.

43. Euclid (*Elements*, chap. 7, def. 5) defines this as a relationship where the minor measures the major. Other definitions agree with each other essentially; see Theon, p. 76; Jordanus, Book 9, prop. 37, 38, 52, 70; Hero of Alexandria, def. 121 and Heath, *Elements*, Vol. 1, pp. 20-24. Boethius' source is Nicomachus, Book 1, chap. 18.

tum. You will name these numbers always with the prefix »sub« added on, as one compared to 2 is a subduplex, one compared to three is subtriplex, one compared to four is subquadruple, and so on.

Since multiplicity and submultiplicity are naturally infinite, their type, through their own generation, will move on in an infinite possiblity. If in the natural disposition of simple numbers through an evenly constituted sequence you were to take equally all even and odd numbers beginning from one, each following the other, they will be duplex, and there is no termination to this kind of speculation. Now natural numbers are placed in this order: 1, 2, 3, 4, 5 , 6, 7, 8, 9, 10, 11, 12, 13, 14, 15, 16, 17, 18, 19, 20. Of these numbers, if you take the first even number, that is 2, it is the duplex of the first, that is of unity; if you take the following even number, that is 4, it is the duplex of the second number, that is, of two. If you take the third even number, that is 6, it is the duplex of the third number in the natural series, that is , of three. If you look at the fourth even number, that is 8, it is the duplex of the fourth number, that is four. This same process proceeds in the remaining numbers into infinity without any stopping.

Triple numbers are born if, in the same natural disposition of numbers, two are regularly skipped over, and the numbers that are after those two are then compared to the natural number, with the third excepted, which, as it is the triple of unity, goes only after two. Now after one and two there is three which is triple one; again after 4 and 5 is 6, which is the triple of the second number, which is two. Again, after, 6, 7, and 8 there is 9, which is the triple of the third number, that is, of three. If someone does this to infinity, he will proceed without any obstacle.

The generation of the quadruple numbers begins if one skips over three numbers. After one, two, and 3 there is 4, which is the quadruple of the first number, that is, unity. Again, if I skip five, six, and seven, the quadruple number eight comes to me, that is after three more numbers, and it is the quadruple of two, or the second number in the natural series. If after 8 I pass three more numbers, that is 9, 10, and 11, then twelve which follows is the quadruple of the third number in the natural order. Now this same progression must necessarily go to infinity and if it always increases by the adding on of terms in the interval, you will marvel to find the ordered ways of the multiplex number. Now if you skip over 4, you find the quintuple, if you skip 5 you find the sextuple, if you skip 6, you find the septuple, and the term of these numbers is always created by the name of the interval of the multiplication minus one. Two skips over one, three skips over 2, the quadruple skips over 3, the quintuple skips over 4,

and thereafter the sequence is according to that order. Every duplex number is even according to the proper sequences of even numbers; if there are
triple numbers, one will always be found even, another odd. The quadruple numbers again always maintain an even quantity and are constituted
by a fourth number after each number. After four, there is 8, six more
than two, then 12, ten more than the binary, and this is the same in the remaining numbers. If there are quintuple numbers, their placing is ordered
according to a similitude of triple with even and odd alternately occurring.

24. Concerning the superparticular number, its types, and their generation.

The superparticular is a number compared to another in such a way
that it has in itself the entire smaller number and a fractional part of it.[44] If
it has half the smaller number, it is called sesquialter, if it has a third it is
called sesquitertius, if it has a fourth, it is called sesquiquartus and if it has a
fifth it is called sesquiquintus. The order of the superparticular number
proceeds with these terms carried on *ad infinitum*. The larger numbers are
called after this manner and the smaller numbers which are contained in
them, their entire sum plus a fractional part, are called subsesquialter, another subsesquitertius, another subsesquiquartus, another subsesquiquin-
tus, and these are extended in this way according to the manner and multitude of the larger numbers. I call the larger numbers »leaders« and the
smaller numbers »followers.«[45]

There is also an infinite multitude of superparticular numbers
because its species is founded on an interminable progression. The sesquialter has leaders after the third number which are naturally triplicate, and
followers after the second number which are naturally even, and they occur in this manner, as the first in the first place, the second in the second
place, the third in the third place and so on. They are described in a long
row against the double and the triple of the natural order in this manner:

1	2	3	4	5	6	7	8	9	10
3	6	9	12	15	18	21	24	27	30
2	4	6	8	10	12	14	16	18	20

44. Nicomachus, Book 1, chap. 19; Martianus Capella, Sec. 761; Jordanus, Book 9, prop. 37,
38, 52.
45. The terms *duces et comites* translated literally are »leaders« and »followers«. Nicomachus used the terms πρόλογος and ὑπόλογος which D'ooge renders as
»antecedent« and »consequent.«

The first row contains the natural number, the second row contains its triple, the third row contains its double. In this diagram, if the third be compared with the second, if the sixth with the fourth, or if the ninth with the sixth or all the triple upper numbers are opposed to the double numbers, a hemiola, that is a sesquialter, proportion is formed. For 3 has 2 in it and half of it, that is one. Six also contains 4 in it and half of it, that is two. Nine encloses 6 in it and half of it as well, that is 3. The process continues in this way with the others.

It should be demanded that if someone wants to consider a second type of superparticular number, that is the sesquitertian—by what rationale would he find it? The definition of such a relationship is: a sesquitertian is that number which when compared has the smaller number once and a third part of it. These are found if all the numbers from 4 are laid out in fours and compared to triplicate numbers laid out from there. The leaders will be quadruples, the followers triples.

Let numbers then be put in order in such a fashion that there are first natural numbers, and below that quadruple, and then triple numbers, with the first triple placed under the first quadruple, the second under the second, the third under the third, and in that manner all the numbers in the first row are arranged in orders against the triple:

1	2	3	4	5	6	7	8	9	10
4	8	12	16	20	24	28	32	36	40
3	6	9	12	15	18	21	24	27	30

So if you compare the first with the first, a sesquitertial ratio is established. Then if you compare 4 to 3, four will have three entirely within itself and a third part of it as well, that is one; and if you compare the second of the top row to the second below it, that is eight to six, you find the same, for eight contains six entirely and a third part of it, that is 2. And through that sequence the same process must go on *ad infinitum.*

It should also be noted that 3 is a follower and 4 is a leader; six is the follower, 8 is the leader. In that same order of the other numbers, in a similar manner, the leaders are called subsesquitertial. The terminology established in this way is able to serve in all the remaining numbers.

25. Concerning the use of knowing the superparticular.

It is admirable and very profound to discover in the ordering of these numbers that the first leader and the first follower are joined to each other without the intermission of another number. In the first we pass from one to the other with no number placed between; in the second, there interposes one, in the third there interposes two, the fourth, three, and so they increase with always one less than the number of their rank between, and this must be found in the sesquialter or sesquitertial or in other superparticular parts. If the fourth were compared to the third, no number would interpose, for in the next place after three there is four. But in order that 6 be put next to 8, namely in the second sesquitertial position, there must be an intermission of one. Between 6 and 8 there is only seven, which is the one number passed over. Again we compare 9 with 12, which two numbers are in the disposition of a third, and a transition of two numbers is made between them: between 9 and 12 there are 10 and 11. In this manner, in the fourth disposition of thirds, there is a fifth, with 4 numbers coming between.

26. A description through which it is shown that the multiplex relationship is anterior to the other types of inequalitities.

Since we have proposed that the multiplex type of inequality occurs naturally and according to its proper following of order, this should have been clear for us in the order of the previous work; but touching on it lightly, again, as we have proposed, we will briefly and plainly give some more instruction.

Let there be such a diagram in which is placed in order all the way to the number ten the natural order of continuous numbers; in the second line, the duplex order should be extended, in the third, the triple order, in the fourth, the quadruple order, and this is done all the way to ten. In this way we will recognize how the types of multiplex number are in a principal position to the superparticular and superpartient. At the same time we will see certain other things: how exact they are in their subtlety, how useful for knowledge, and how delightful for exercise.

LONGITUDE

L	1	2	3	4	5	6	7	8	9	10	L
A	2	4	6	8	10	12	14	16	18	20	A
T	3	6	9	12	15	18	21	24	27	30	T
I	4	8	12	16	20	24	28	32	36	40	I
T	5	10	15	20	25	30	35	40	45	50	T
U	6	12	18	24	30	36	42	48	54	60	U
D	7	14	21	28	35	42	49	56	63	70	D
E	8	16	24	32	40	48	56	64	72	80	E
	9	18	27	36	45	54	63	72	81	90	
	10	20	30	40	50	60	70	80	90	100	

LONGITUDE

27. The reason for and the explanation of the above formula.

If the sides of the proposed diagram are examined and those numbers are compared which make angles from one to ten on the top and proceed by ten below, and the lower orders are compared to these, that is beginning from 4, and they are put each next to the neighboring term in the second rank, the duplex order then appears, and that is the first type of the multiplex. It is shown thus, that the first number exceeds the first only by a unity, as two exceeds one; in the second rank the second number exceeds the second by two, as four is 2 more than 2, the third exceeds the third by three, as six exceeds three, the fourth exceeds the fourth by a magnitude of four, as 8 is more than four, and through the same sequence the numbers in the second row show themselves higher in quantity than the smaller numbers. If we examine the third angle, which beginning from 9 extends to the latitude and longitude on either side of the number three, and if we compare this number with the first number in latitude and longitude, the third type of multiplex appears. So, if we make this comparison through the number 10, each number will exceed the next connected to it according to a naturally made connection of equality. The first exceeds the first by two, as three exceeds one; the second exceeds the second by four, as six exceeds two, and the third exceeds the third by six, as nine exceeds three, and the rest of the numbers increase according to the same mode of progression. This is because the natural integrity of numbers puts this fact before us and there is nothing but sheer computation just as it appears to us in this descriptive diagram.

If one should wish to compare the term of the fourth angle which is noted in the quantity of the number 16, and determines the number 4 in longitude and latitude with a proportion compared through the number 10, he would note the multitude of a quadruple, and on this ordinal number there is a progression over itself. So, the first surpasses the first by three as 4 surpasses unity, the second surpasses the second by six, as 8 surpasses two, the third surpasses the third by nine, as twelve surpasses three, and the following sums of the three rank surpass each other always with the adjacent quantity added on.

If someone looks at the lower angles, he will come in the same way through all the types of the multiplex, through a most disposed ordering, to ten. If one looks for a type of the superparticular in this description, he will find it in this way. If he takes note of the second angle, whose beginning is four, and he goes two beyond that and adopts the following order, a sesquialter proportion is defined. The third item of the second rank is a sesquialter, as 3 to 2, or 6 to 4, or 9 to 6, or 12 to 8, and likewise in the other numbers which are in the same series of numbers; if such a conjunction is put together, no change of this relationship will disrupt it.

This same progression of numbers exists also in the duplex order. The first number exceeds the first number, that is the three exceeds the two, by one; the second exceeds the second by two, the third exceeds the third by three and so on in succession. Now if the fourth order is compared to the third order, a sesquitertial comparsion will be assembled, as 4 to 3, and if you put together the rest in the same order, there are 4 to 3, 8 to 6, and 12 to 9. Do you see how in all these numbers the sesquitertial comparison is kept? After these, going to the numbers below them, if you do the same thing, and you compare the following against each other, you will find all the types of the superparticular without any hindrance.

Such is the divine nature of things in this disposition that all the angles are tetragons.[46] Now I will define a tetragon as briefly as I can explain it here, since it will be discussed later more fully. A tetragon is that which two equal numbers achieve when multiplied together, as is also found in the above scheme. One times one is one, and this is a tetragon in potency. Then two times two are four; three times three are nine. The multiplication of numbers in the first order with themselves will always bring this [a tetragon] about. Around these numbers, that is, around the angles, there are the longilateral numbers.[47] I call those numbers longilateral [or longer

46. Boethius here apparently limits the term »tetragon« to indicate square numbers.
47. Nicomachus called the longilateral number heteromecic   (ἐτερομήκεις)   On the

by one side] which multiply themselves in exceeding each other by one. Now around 4 are 2 and six. Two is born from one and two, when you multiply one twice. Unity is preceded by unity in two. Six comes from two and three, since twice three gives six. The numbers 6 and 12 inclose the number nine, from which three and four are born: three and four multiplied make 12. Six comes from two and three, because two times three makes six. These numbers are all created from the major sides.

Now since 6 is born of two and three, three surpasses the number two by one, and all the other numbers are of such a type that they are created from the terms of the first and second order multiplied with each other. Thus, what is born from two longilateral lines placed next to each other is a tetragon. And again from two tetragons adjoining and twice the number longer by one side always comes a tetragon.

And as the first unity of the whole scheme of angles refers to the tetragonal angles of the positive numbers of one angle, so for the other which is contrary to it, the third, there are two angles which have the second unities. And the two angles make the equal sides of the tetragonal angles, because they are contained under them and are made from one of them which is opposite it. [Boethius seems to be saying that as the first of the angles of the entire diagram, that is unity, is a tetragon, so is the angle opposite it, 100, a tetragon. The other two angles, 10 and 10, when multiplied, are equal to the first and the third multiplied and, like them, they make a tetragon.]

There are many other admirable and useful things which we are able to draw from this description but we have allowed them to go unmentioned because of censuring brevity. Let us now turn to the following items of our exposition.

28. Concerning the third type of inequality, which is called superpartient, and its types, and their generation.

After these first two types of proportion, the multiplex and the superparticular, and the types which are under them, the submultiplex and the subsuperparticular, there is found a third type of inequality which we have already called the superpartient.[48] This occurs when one number

diagram, chap. 26, six is the heteromecic or longilateral number between 4 and 9. Boethius also calls such numbers »longer by one side,« *numeri altera parte longiores.* The relationship described at the end of this chapter shows that the sum of two successive squares and twice the longilateral number between is always a square.

48. Nichomachus, Book 1, chap. 20; Martianus Capella, Sec. 762; Jordanus Book 9, prop. 42, 53, 54.

compared to another contains that number entirely within itself and its aliquant parts as well, either two or three, or 4, or however many that comparison brings out. This relation begins with two times three parts. Now if a number has two halves which it contains entirely in itself, that double is put together as a superpartient. It will have either two fives or two sevens or two nines and so in progression. If it has in addition two single parts of the smaller number, in those parts the larger term exceeds the smaller term by odd numbers. Now if it contains that total and two quarters of its parts, it would necessarily be found to be superparticular, for two quarters is a half and this is a sesquialter comparison. If two sixes occur, again it is of the superparticular, for two sixes is a third part; if the two parts were to be compared, they would result in a form of the sesquitertial relationship.

The followers are generated next and are called subsuperpartient. These are numbers considered in terms of another number, plus two or three or 4 or whatever more of its fractional parts. If, therefore, a number contains another number in itself and two of its fractional parts, it is called superbipartient; if it has three fractional parts, it is called supertripartient; if it has four, it is called superquadripartient. So, one may work out all these terms and proceed *ad infinitum*.

The order of these numbers is natural insofar as from three, all even and odd numbers are disposed and then under them there are prepared other numbers which, beginning from five, are all odd. With these numbers so disposed, if the first is compared with the first, the second with the second, the third with the third, and the rest carefully compared with the others, the relation of the superpartient is created. Let these numbers be put down in the following manner:[49]

3	4	5	6	7	8	9	10
5	7	9	11	12	15	17	19

If we were to consider 5 in comparison to the number three, it would be a superpartient, which would be called superbipartient, for the five would have the entire three in itself and two fractional parts of it. If this kind of speculation is carried to the second order of numbers, the supertripartient proportion will be recognized, and so on with the following numbers through all the disposed numbers *ad infinitum*. You will thus find the assembled and ordered types of this numerical relationship.

Such is the method of finding how the individual numbers are created all the way to infinity, if one cares to know this. The relationship of the

49. In Friedlein's text, the first number mistakenly reads as IIII instead of III.

superbipartient is such that if it is doubled in both terms, the proportion of the superbipartient is always created. If someone were to double 5, he would make 10; if he doubles 3, he makes six. If someone compares 10 against six, the compared numbers make a superbipartient relationship. If you double these again, the same order of proportion grows from it, and if you do the same infinitely, the status of this relationship will not change. If you try to find the supertripartient, the first supertripartient is 7 to 4, and if you triple it, the rest emerge. The numbers thus born which you produced by the multiplication of three may again bring about the same relationship. If the superquadripartient were to progress *ad infinitum*, one should add the first roots that you multiply by fourths, that is 9 and 5, and by that multiplication the series will be produced, again by fourths, and you will find that the proportion will grow surely with unabated logic. The other types grow always by one multiplication more from the developing root numbers. I call these root numbers of proportions described in the above disposition as though it were from them that every sum of the said comparison begins its growth.

And this must be observed that when two parts more from a smaller number are in a larger number and they reside there in terms of a third, then this is called a superbipartient, since it has the third part of a smaller number, a third superbipartient. When I say supertripartient, it is necessary to understand the fourth supertripartient, because the three is always exceeded by a fourth part. The superquadripartient is always understood as the fifth superquadripartient, and according to that mode, in the rest of the numbers the understood part is created over the existing part with one always added on, so that their order and development are in these terms: the number called superbipartient is at the same time called superbitertient. The one called supertripartient is also called supertriquartus. And the one called superquadripartient is called superquadriquintus and with this same order the terms are produced *ad infinitum*.

29. Concerning the multiplex superparticular.

Now these have been the first and simple types of one quantity related to another. There are two other types which are thought of as put together from other principles, such as the multiplex superparticular and the multiplex superpartient;[50] their followers are the submultiplex superparti-

---

50. Nicomachus, Book 1, chap. 22; Martianus Capella, Sec. 763-764; Jordanus, Book 9, prop. 37.

cular and the submultiplex superpartient. In these, as in the aforesaid proportions, smaller numbers are all said to be added under a proposition whose definition can be given thus: the multiplex superparticular is one that as often as one number is compared to another number, it has that number more than once and then a part of it, that is, it has the number double, or triple, or quadruple, or however many times, and then a certain fraction of it, either a half or a third, or a fourth, or whatever other part might be added to it by excess. This number therefore consists of a multiplex and a superparticular. Because it has the number compared to more than once, it is multiplex. Because it exceeds the smaller number in having a fraction of it, it is superparticular. Thus from each name is the term made up and its types are made up according to a notion of those proportions from which that number takes its origin. Now that first part of this term, which holds the name of multiplex, must be denoted by the term of the multiplex number of the species. The superparticular part is named from the term by which the species of the superparticular number is called. The duplex sesquialter is said to be that which has another number twice and its half; that which has a third is a duplex sesquitertius; a fourth, is a duplex sesquiquartus, and so on. If one number contains another number three times entirely then a half of it, or a third, or a fourth, it is called triplex sesquialter, triplex sesquitertius, triplex sesquiquartus, and in the same manner do we proceed in the remaining numbers. The number which has the smaller term four times is a quadruplex sesquialter, quadruplex sesquitertius, quadruplex sesquiquartus. And however often it contains the total number in itself it will be called in terms of that species of the multiplex number, and whatever fraction of the compared number it includes, it is called according to a superparticular comparison and relationship.

Here are some examples of these relationships. The duplex sesquialter is 5 to 2. Five has the two twice and then half of it, which is one. The duplex sesquitertius is seven compared to three. Nine compared to four is a duplex sesquiquartus. If we compare 11 to 5, we have a duplex sesquiquintus. These proportions are always generated from numbers disposed into an order from 2, all naturally in even and odd terms. If opposed to each number in this order from the number five you compare the odd number, diligently and carefully putting first by first, second by second, third by third, this sort of scheme emerges:[51]

51. These schemata are adapted from Nichomachus.

2	3	4	5	6	7	8	9	10	11
5	7	9	11	13	15	17	19	21	23

If you put down even numbers from the number two and under them, beginning from five and skipping over each term by five, you compare them to the terms above, then all the duplex sesquialter proportions are created, as in the following description:

2	4	6	8	10	12
5	10	15	20	25	30

If the disposition of numbers begins with three and skips over each by three in the natural order, and you compare these to numbers which begin with seven and pass over each seventh term, all the duplex sesquitertian numbers are generated from a diligently arranged comparison, as the diagram below illustrates:

3	6	9	12	15	18	21
7	14	21	28	35	42	49

If you put the quadruple numbers in order, that is the quadruple order of natural numbers, as the quadruple of unity, two, three, four, five and of those following them, and to these you compare terms beginning from the number 9 and always proceed by nine, then a diagram of the duplex sesquitertial proportion is built up:

4	8	12	16	20	24
9	18	27	36	45	54

The type of number called the triplex sesquialter proceeds in this manner: all the numbers are disposed from the number 2 by even numbers in order, and to them are compared numbers beginning from 7 going over each by seven in the accustomed manner of comparing each to the other:

2	4	6	8
7	14	21	28

If beginning from the number 3 we dispose all the triple numbers of the natural order, and if we compare to them numbers from ten and adding ten, there will emerge all the triple sesquitertial numbers in this sequence of terms:

3	6	9
10	20	30

30. Concerning examples of the multiplex superparticular and how to find them in the above diagram.

Of these and of those which follow, we are able to make out examples fully and clearly in the first description which we gave, where we spoke of the superparticular and the multiplex and where the sums are put in multiples from one all the way to ten.[52] Then when all those which follow are put together with the numbers of the first rank, they give the ordered and fitting types of the multiplex. If you compare to the second rank all the numbers which are of the third rank, you will recognize the ordered types of the superparticular. If to the third order you compare those in the fifth rank, you will see conveniently placed the types of the superpartient numbers. The multiplex superparticular number is shown when to the second rank are compared all the numbers which are in the fifth order, or those which are in the seventh, or those which are in the ninth. So if there is such a description *ad infinitum*, the types of these proportions will be created *ad infinitum*. It is also obvious that the followers of these numbers are always called with the prefix »sub«, as in the subduplex sesquialter, subduplex sesquitertius, subduplex sesquiquartus and the rest in this manner.

31. Concerning the multiplex superpartient.

The multiplex superpartient[53] is that number which as often as it is compared to another number, contains the other number in itself entirely more than once and one or two or more numbers in addition, according to the form of the superpartient number. Now in this number, on account of the reasons given above, there are not 2 halves nor 2 fourths nor 2 sixths, but 2 thirds or 2 fifths or 2 sevenths according to a similar sequence of prior numbers. It is not difficult according to the examples of the prior numbers given to find these numbers and the numbers outside of our examples. These will be called according to their own fractions, duplex superbipartient, or duplex supertripartient, or duplex superquadripartient and again, triple superbipartient, and triple supertripartient and triple superquadripartient and so on, as 7 compared to 3 makes a duplex superbipartient and 16 compared to 6 and all the numbers which, beginning from 8, with eight numbers skipped over, are compared to those which begin from 3 and proceed over every third term. For those who are diligent, it will not be difficult to find other fractions according to the aforesaid manner for this number. Here we must remember that the smaller numbers and followers are not named without the prefix »sub« as are the subduplex superbipartient and the subduplex supertripartient.

52. See the diagram in Book 1, chap. 26
53. See Theon, pp. 127-29.

32. A demonstration of how every inequality proceeds from equality.

Now it remains for us to treat of a certain very profound discipline[54] which pertains with sublime logic to every force of nature and to the very integrity of things. There is a great fruitfulness in this knowledge, if one does not neglect it, because it is goodness itself defined which thus comes to knowable form, imitable by the mind. By means of it, primary nature and the infinite ugliness of evil are perceptible through the constant propriety of its substance; it rests on none of its own principles but always by nature is derived from the definition of good principle as though put together by the impress of a good form in some sign, and is saved from the flux and change of error. The mind holds in check excessive cupidity and the immoderate frenzy of anger like one who rules, strengthened by this pure knowledge; this knowledge, tempered by goodness, establishes the forms in inequality. This matter will become obvious if we understand that every type of inequality arises from a prior equality so that equality is itself as it were the matrix and, taking the force of a root, it gives depth to the types and orders of inequality.

Let there be put down for us three equal terms, that is three unities, or three twos, or however many you want to put down. Whatever happens in one, happens in the others. From these numbers, then, according to the order of our precept, you will see born first the multiplex and in these the duplex first, then the triplex, then the quadruples and so on according to that order. Again, if the multiplex numbers are reversed, from these the superparticular emerge, and from the doubles come the sesquialters, from the triples come the sesquitertient, from the quadruples come the sesquiquartant, and the rest according to this manner. From the superparticular inverted it is necessary that the superpartient are born, and from the sesquiquartant the superquadripartient proceeds. With the prior superparticular placed rightly and not converted, the multiplex superparticular arises; with the superpartient rightly placed, the multiplex superpartient numbers are produced. Now these are the three rules: that you make the first number equal to the first, then put down a number equal to the first and the second, finally one equal to the first, twice the second, and the third. When you do this in equal terms, the numbers which arise from the process will be duplex; and if you do the same with the duplex terms,

54. Boethius here presents the theory of the »three rules«; see Cantor, Vol. I, p. 432 and Nesselmann, p. 198. The notion that equality is more elementary than inequality has a long tradition in Greek number theory; see Theon, p. 135, where he attributes the notion to Eratosthenes.

the triples are brought out, and from these the quadruple, and this will explicate all the forms of the multiplex number *ad infinitum*. Therefore, let these terms be put down:

1	1	1

Let there then be put down the first number equal to the first, that is one; next a number equal to the first and the second, that is 2; then one equal to the first, double the second, and the third, that is one, twice one, and one, which make 4, as in this diagram:

1	1	1
1	2	4

Do you see how the sequent order in duplex proportion is constructed? Now do the same again from the duplex order, so that the first is equal to the first, that is to one, the second to the first and the second, that is to one and two, which is three, then to the first, which is one, twice the second, which is four, and the third, which is four, and together this makes nine and this formula emerges:

1	1	1
1	2	4
1	3	9

If you do the same again from triple numbers, a quadruple continuity will be created. Let the first be equal to the first, that is one, then the next equal to the first and the second, that is 4, and the final equal to the first, double the second, and the third, that is 16:

1	1	1
1	2	4
1	3	9
1	4	16

In the rest we may use the precept of three numbers according to this formula.

If the numbers born from equals are multiplex, let us set them out and reverse them according to these rules, so that they are turned about in order; the sesquialter is created from the duplex, the sesquitertius from the triple, and the sesquiquartus from the quadruple. Let there be three duplex terms, which are created from equal numbers, and that which is last, let it be put first in this manner:

4	2	1

Then let it be set up in this order, that the first is equal to the first, that is 4, the second to the first and second, that is 6, and the third equal with the first, twice the second, then the third, that is 9:

4	2	1
4	6	9

Then behold there will arise for you a sesquialter quantity from this duplex order of terms.

Let us now see, according to the same manner, what is born of the triple order. The triple order should be set down as above:

9	3	1

This order has been reversed, as was the duplex. Then let there be put down a number equal to the first, that is 9, then a number equal to the first and the second, that is 12, then one equal to the first, twice the second, and the third, that is 16.

9	3	1
9	12	16

Again, the second type of the superparticular number, that is the sesquitertius, is created. If someone wishes to do the same with the quadruple order, the sesquiquartus is born, as the following diagram shows:

16	4	1
16	20	25

If someone does the same with all the parts of the multiplex number into infinity, he will suitably find the order of the superparticular. But if someone should convert the reversed superparticulars according to these rules, he would see the superpartient growing continuously, and see that from the sesquialter the superbipartient is created, and that from the sesquitertian the supertripartient is created, and that the others are all born according to the common types of denomination without an interruption of their order. They would be put down thus:

9	6	4

Of the above description, therefore, first a number equal to the first should be written, that is 9, second there should be added a number equal to the first and the second, that is 15, then to the first, twice the second, and the third, which is 25:

9	6	4
9	15	25

If we turn in the same manner to the sesquitertian, the superpartient order will be found. Let there be the first position of the sesquitertian:

16	12	9

Next, according to the above manner there is placed one equal to the first, that is 16, next one equal to the first and the second, that is 28, then one equal to the first, twice the second, and the third, that is 49. The entire sum of these numbers disposed gives the supertripartient:

16	12	9
16	28	49

Again, if you reverse the sesquiquartan in the same way, immediately the superquadripartient quantity will be produced, as in the form which you see put here:

25	20	16
25	45	81

It remains for us to show how from the superparticular and superpartient the multiplex superparticular or the multiplex superpartient are born, of which I will make only two diagrams. If we place the sesquialter in proper, not reversed order, the duplex superparticular grows of it. Let it be put in this way:

4	6	9

Then we place, according to the above method, a number equal to the first, that is 4, a second equal to the first and the second, that is 10, then one equal to the first twice the second and the third, that is 25:

4	6	9
4	10	25

So thus is the sum of the duplex sesquialter produced. If we put down the sesquitertial numbers, not reversed, but in proper order, the duplex sesquitertial is found, as the following diagram shows:

9	12	16
9	21	49

If we turn our attention to the superpartient and we dispose the numbers in an orderly way according to the above rules, we will find the multiplex superpartient produced in an orderly fashion. This is the formula of the superpartient properly disposed:

9	15	25

Then there should be written a number equal to the first, that is 9, then one equal to the first and the second, that is 24, finally one equal to the first, double the second, plus the third, that is 64:

9	15	25
9	24	64

Do you see how the duplex superbipartient comes from the superbipartient? If I put down the supertripartient, without doubt the duplex supertripartient will be found, as it is seen in the following description:

16	28	49
16	44	121

And so therefore from the superparticular or the superpartient, the multiplex superparticular, and the multiplex superpartient originate. And so it is established that equality is the principle of all inequalities. Then from inequality all other things are derived.

We believe we could have spoken more about these things, but we should not pursue infinite matters, nor should we linger over very obscure things which hold back our minds when we would proceed to more useful matters.

The end of the First Book

Here begin the Chapter Titles of the Second Book

Here begins the *Second Book.*

1. How every inequality is reduced to equality.

It has been briefly stated in the exposition of the first book how every substance of an inequality is preceded by equality from the principle of its generation.[1] But let us inquire about what the elements of things may be, that is, elements from which all things are put together and again into which all things made are dissolved. These elements may be considered as the utterings of an articulate voice and from them emerges the progressive joining together of syllables and into their terms it may again be resolved. Sound holds a similar force in musical compositions.

We are not ignorant of the fact that four elements make up the world; as they say, material bodies are born from fire, air, earth, and water.[2] But for these four elements, again, there is a prior composition, so that as from the very source of equality and inequality we may see every species come forth, and we may resolve every species of inequality into equality as though to a certain element of its own origin.

Now this matter may be broken down under a triple arrangement,[3] and that art of resolving may come to any three, unequal but proportionally disposed, that is so that each holds the same median to the force of the proportion as the one which is extreme holds to the median, in whatever type of inequality it may be, whether in multiplex, superparticular, or superpartient or in others created from these, the multiplex superparticular or the multiplex superpartient. In all, the same and undoubted ratio will be maintained. So, as it was said, with three terms set up in equal proportion, we always draw the last one to the median and with that last one we always put the first term and with it put the second term which is left from the median. For the third of the proposed terms, let us take the first term once, twice the second, that is which is left from the median, plus that which is left from the previous third sum, and this we constitute as the next third term. You will see when this is done that the sums are changed into a smaller mode and the comparisons and proportions are reduced to a more basic condition, so that if it is a quadruple proportion, it

1. Nicomachus, Book 2, chap. 1
2. For the place of these elements in Greek natural philosophy, see John Burnet, *Greek Philosophy* (New York, Macmillan, 1914), vol. 1, p. 26. They subsequently became very important to medieval thought as well; see A. C. Crombie, *Medieval and Early Modern Science* (New York, Doubleday, 1959), vol. 1, pp. 129-133.
3. Nicomachus, Book 2, chap. 2.

goes to a triple, then to a double, and from these finally to equality. If it is a superparticular sesquiquartus, first it goes to a sesquitertian and finally to three equal terms. We will show this in an example for the multiplex proportion only and the same logic of rules will help a skilled experimenter as well in other types of inequality. Let three terms, each the quadruple of the previous, be set down:

$$8 \qquad\qquad 32 \qquad\qquad 128$$

Then subtract the smaller one from the middle term, that is 8 from 32, and 24 is left. Put down the first term of eight, then in second place put what remains of the middle term, that is 24, and so we have two terms, 8 and 24. Then from the third term, which is 128, take the first, that is 8, and twice the second, which is twice 24, and we have 72. Now when we have put these terms down, from a quadruple equality we see that a triple proportion is derived. These terms are: 8, 24, 72. If you do the same thing to these, the comparison will go down again to a duplex. In the first place, put a term equal to the smaller number, that is 8; from the second take the first, and 16 is left. Then from the third, that is 72, take the first, which is 8, and double the second, which is 16, and the remaining portion will be 32. When you have put these figures down, the relationship is reduced to a double proportion: 8, 16, 32. If the same is done again with these numbers, we will reduce the whole set to the sums of equality. Put down a term first equal to the small number, that is 8; then from 16 take 8, and 8 remain. Then from the third, that is 32, subtract the first, that is 8, and twice the second, that is twice 8, and there then remains 8. When you have put this down, the first equality will occur for us, as the following terms show:

$$8 \qquad\qquad 8 \qquad\qquad 8$$

If one directs his attention from here to the other types of inequality he will unwaveringly find the same accord. For that reason it must be said, and not doubted with any apprehension, that as unity is the substance and principle of any constant quantity, so equality is the mother of any quantity related to any other thing. We have demonstrated that from equality comes the first procreation of a relationship and to it again is its final resolution.

2. Concerning the discovery in each number of how many terms of the same proportions are able to precede it, with their description, and an explanation of the description.

There is in this matter a certain profound and marvelous speculation and, as Nicomachus says in explaining the »mystical theory,« something which relates both to the Platonic generation of the soul in the *Timaeus*[4] and to the intervals of harmonic discipline. In that book we are instructed to produce and extend three or four sesquialter or any other number of sesquitertial proportions and sesquiquartal comparisons. We are told to extend them according to a proposed order, always continually. Lest, however, one become overwhelmed by this labor, which is great indeed, as is usually the case, we should trace out by reason in how many numbers and how many superparticulars this can be done.

All multiplex numbers are principle of so many proportions similar to themselves as is any one place distant from unity.[5] When I say similar to themselves, it is such that the multiplex of the duple, as it was arranged above, creates a sesquialter and the leader is a triplex of the sesquitertial and the quadruple of the sesquiquartus. The first duplex will have only one sesquialter, the second two, the third three, the fourth four, and according to this order the same progression is carried out to infinity, nor can it ever happen that the number of proportions will be smaller than strict equality with the place from unity. The first duplex number is a binary number, which accepts only one sesquialter relationship, that is a ternary, for a binary compared to a ternary produces a sesquialter proportion. But a ternary number cannot be divided in half; there is no other number to which it may be compared in a sesquialter ratio. A quarternary number is twice duplex and therefore it precedes two sesquialters. Then the number six compared to itself, or to six, and half again added on makes nine; so these are sesquialters, 6 compared to 4 or 9 compared to 6. Now 9, since it lacks a half, will be excluded from this comparison. A third duplex is 8. This comes before three sesquialters. The number 12 is compared to it, then 18 to twelve, then 18 to 27. But 27 is unable to be made into halves. The same must happen in the following numbers as we have given below in proper order. This must always occur as by a certain divine, not human, divising, that as often as the final number is found, the one which is equal to the place of the double number from unity is such that it cannot be divided or separated into equals.

1	2	4	8	16	32
	3	6	12	24	48
		9	18	36	72
			27	54	108
				81	162
					243

4. Plato, *Timaeus*, sec. 34-35.
5. Nicomachus, Book 2, chap. 3.

The same ratios hold true even in triple numbers. From them the sesquitertians are created. Since the first triple is a ternary number, it has one sesquitertian, that is 4, and of this 4 a third part cannot be formed; so this number lacks the epitrita. The second number, which is nine, does have twelve as a sesquitertian to itself, because that does have a third part; then there is compared to 12 the number 16, which is [not] able to be divided into sections of a third part.[6] Now 27, which is the third triple, has 36 in a sesquitertial relationship to it, and this again is compared to 48 in the same proportion. When 48 is put down, it would terminate the development of that proportion because to 48 you cannot match any sesquitertial relations since triple parts are not contained in it. And this is found in all triple numbers, that the last number of that proportion has as many numbers preceding it as the first stands distant from unity; however many numbers it has of that proportion beyond itself so many does the first of them lie from unity. That part of it cannot be found which, when a number is compared to it, is not able to make the same proportion. This is a diagram of triple proportions:[7]

1	3	9	27	81	243
	4	12	36	108	324
		16	48	144	432
			64	192	576
				256	768
					1024

A diagram of the quadruple numbers is made according to the following form; if one approaches instructed in the previous principles, he should by no reason lack confidence in these matters. Concerning the remaining multiplex numbers, you will note the same regularities:

1	4	16	64	256	1024
	5	20	80	320	1280
		25	100	400	1600
			125	500	2000
				625	2500
					3125

6. It seems that Boethius' text has a negative missing, which I add in brackets: "Numerus XVI, qui tertiae partis sectione [non] solutus est." All the texts which Friedlein examined and all which I saw had the same omission. Nicomachus has: "The second triple is 9 and hence will begin a series of only two sesquitertian ratios, 12, its own and 16, that of 12; but 16 cute off further progress for it is not divisible by 3 and hence will not have a sesquitertian." (D'ooge, p. 233).
7. Nicomachus, Book 2, chap. 4.

This may also be seen in the case of the superparticular as we have just shown. All the superparticulars may be created in order, beginning first with the multiplex numbers, then the duplex sesquialters, then the triple sesquialters. In these numbers it is a marvelous thing. Now where the first rank was duplex, under each of those, numbers are continuously placed in alternate places which according to the series of the latitude are duplex. If there are triple numbers in the first rank, the lower orders are also increased in themselves by a triple multiplication in their terms. Also in the quadruple order the quadruple numbers in their infinite mode of speculation will not fail to be extended quadrupally. It is necessary that the angles[8] of all the numbers be multiplex. Of the duplex numbers they are triple; of the triple numbers, they are quadruple; of quadruple numbers, they are quincuple, and all the numbers fall together with each other to the same unchangeable logic of order. Since we have finished explaining these things, our inquiring analysis will be turned to the following series of work.

3. What multiplex interval comes from what superparticular, with what interval placed between, and the rules for finding this.

If the first two types of the superparticular are put together,[9] the first species of multiplication arises. Every duplex is put together from a sesquialter and a sesquitertian, and every sesquialter and sesquitertian is joined together to form a duplex. The sesquialter is three to two, the sesquitertial is four to three, the duplex is four to two.[10]

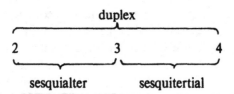

So thus, the sesquialter and sesquitertial put together form a duplex relationship. But if the duplex were divided between the double number and the mid point, a half is found which to the one extreme is a sesquialter and to the other is a sesquitertial. With six and three being placed opposite

8. That is, the numbers in a line running diagonally from one to the highest.
9. Nicomachus, Book 2, chap. 5.
10. In the following diagram, Friedlein erroneously has III for IIII (p. 84).

each other, they form a duplex; if a median of 4 is put in the middle between 3 and 6, to the three it would have a sesquitertian relationship, to the 6 it would be in a sesquialter relationship.[11]

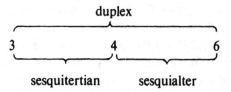

Rightly is it therefore said that a duplex is joined together from a sesquialter and a sesquitertian, and these two types of superparticular create a duplex; that is the first type of multiplex quantity. Again from the first type of multiplex, that is from the duplex, and the first of the superparticular, that is the sesquialter, there is joined a triplex relationship, which is another multiplex. Twelve is double the number 6; 18 is a sesquialter to 12 and a triple to 6:

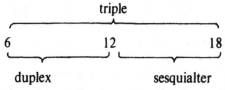

Now if these same 6 and 18 are put together and the number 9 is put in the middle, it will be sesquialter to 6 and a duplex to 18. The 18 will still be a triple to 6.

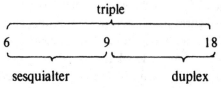

So from a duplex and a sesquialter arises the ratio of a triple proportion, and into these relationships a triple can again be resolved. If this triple number, which is the second type of multiplex, is compared to the second type of superparticular, thereupon will be formed the quadruple. And into

11. The following diagram is not found in Nicomachus.

those components again will the second mode be resolved by a natural division, which we have shown above. Now if a quadruple joins itself to a sesquiquartus, the quincuple will thereupon come about; and if the quincuple is joined with the sesquiquintus, the proportion of the sescuple will immediately be put together. So thus according to this progression will all the types of the multiplex be born without any change of order, so that the duplex with the sesquialter creates the triple, the triple with the sesquitertian creates the quadruple, the quadruple with the sesquiquintus creates the quincuple and in the same way it proceeds and no termination may impede this continuation.

4. Concerning quantity constant of itself, which is considered in geometric figures and in which is a common element of all magnitudes.

We have been speculating about those things which concern quantity in relation to another, and let the things said until now suffice. In the following discourse, let me treat of some ideas concerning that quantity which consists of itself;[12] it is not referred to anything else, and this may be of some profit for us when afterwards we treat again of quantity in terms of another method. Mathematical speculation loves to be established by alternate types of proof. We will now begin to take up a discussion concerning those numbers which deal with geometric figures, their space, and dimensions, that is, concerning linear numbers, triangles, quadrangles and others which extend only on a plane dimension as well as figures joined by an unequal composition of sides, that is solids, such as cubes, spheres, and pyramids, and scalene or irregular and cubic figures, all of which are the proper consideration of geometry.[13] But as the science of geometry is produced from arithmetic as from a root or mother, so are the the seeds of those figures found in elemental numbers, and if we make a plane, it would assume all these intermediate disciplines and it would make firm these established items in detail. It must be recognized that these signs of numbers which are put down and which men describe in designation of numbers were not formed by natural institution. Nature does not arrange the figure of five, »5,« or of ten, that is »10,« and other figures of this type which we write; rather, custom fixes them. Put together five ones or ten or however many other which men wish to denote with these figures, unless one should wish to designate unities, and so often let him put down marks.

---

12. Nicomachus, Book 2, chap. 6.
13. See below, chap. 7-49. Concerning the relationships of these disciplines, see also Book 1, chap. 1.

So we, however often we wish to demonstrate something, especially in these formulas, will not be wearied to put down a multitude of ordered figures. Thus when we wish to demonstrate five, we will make 5 strokes, and lay them out in this way: lllll. And when we wish to make seven, likewise, and when ten, no less, because it is very natural that whatever the number, so many elements it contains in itself and so many designations of unity are assigned to it.

Therefore unity has the potential of a point, the beginning of interval and longitude; it is not itself capable of interval or longitude, just as the point is the beginning of the line and the interval, although it is itself neither interval nor line. Nor does a point put upon a point bring about an interval, any more than if you joined nothing to nothing. It is nothing and nothing comes from nothing. The same proportionality exists between equalities. Now if there were equal terms, so much it is from the first to the second as from the second to the third, and between first and second, or second and third, there is no quantity of interval or space. For if you put down three sixes in this manner,

6   6   6

as the first is to the second, so is the second to the third, and between the first and the second there is no difference: no interval of space distinguishes six and six. So, unity multiplied by itself begets nothing. Likewise unity begets nothing from itself except itself. Because it lacks an interval, it does not have the power of generating intervals, which is not seen to happen in other numbers. Every number multiplied by itself results in another number larger than itself, because the multiplied intervals extend themselves by the greater distance of the space. That number which exists without an interval does not have the power of extending beyond what it is itself. So from the principle, that is from unity, the first longitude of all things grows, which by the principle of the binary number uncoils itself toward all numbers because the first interval is a line. Two intervals are longitude and latitude, that is line and surface. Three intervals are longitude, latitude, depth, that is the line, the plane and the solid. Aside from these intervals none other can be found. Either there is one interval which is longitude, or something is extended in two intervals as when something has latitude and longitude, or it is enclosed in a triple dimension of intervals, if it is reckoned in latitude, longitude and depth. Beyond these, nothing is able to be found, so thus the forms of the six motions may be put with the natures of intervals and numbers. One interval contains two motions in itself, so that in the three intervals the sum of six motions are contained in this manner: in longitude, before and behind; in latitude, left and right; in depth, front and back[14]. It is necessary that whatever solid body may exist, it should have length, width, and depth, and whatever contains these three in itself, that thing is by its very name called a solid. These three things are concerned with each other by an inseparable connection, in every body, and it has been so constituted in the nature of bodies. If anything should be lacking in one of these dimensions, that body is not solid. That which maintains only two intervals is called a surface; every surface is contained only by width and breadth. In this matter the converse also holds; everything which is a surface maintains a length and a breadth, and that which retains these dimensions is a surface. This surface is distinguished from a solid body only by one dimension, which again exeeds the line by one dimension, which has always been accustomed to maintain the nature of length and breadth. The line, to which has been attributed the nature of one dimension, is exceeded by the surface in one dimension and is exceeded by the solid in two dimensions. The point is exceeded by the line in another dimension, that which remains, length. If a point is superceded by one dimension in a line, it is exceeded in a surface by 2 dimensions; a point is removed from solidity by 3 dimensions of intervals, and so it is that a point exists without magnitude or a body or dimension of an interval. It is bereft of length, width, and depth. It is the principle of all intervals and indivisible by nature, and the Greeks call it *atom*; it is so diminished and very small that parts of it can not be found. Therefore the point is the principle of the first interval, but it is not an interval; it is the head of the line, but not yet a line.[15] It is the principle of a line, just as the line is the principle for a surface. It is not the surface, but the head of that second dimension, and it does not include the second dimension. It also falls into the understanding of the surface since it naturally gives beginning to a solid body and a triple dimensional thing. It is itself not extended by a triple dimension of interval nor is it solidified by any density.

---

14. The six categories of relative position are widely cited among Greek philosophers. See Plato, *Timaeus*, Sec. 43 a-b, where he adds rotation and Philo Judadeus, *De Opificio Mundi*, Sec. 122; Macrobius, *Comentary on the Dream of Scipio*, Book 1, chap. 6; Pseudo-Iamblichus, Sec. 55; Martianus Capella, Sec. 736.
15. Nicomachus, Book 2, chap. 7.

5. Concerning linear number.

Such is unity in number: it is not itself a linear number, but it is the principle for extending a number into length, since it is lacking all width yet it is the beginning of extending a number into the dimension of width. Plane numbers also, though they are not solid body, are still the head with an added width of solid body. This will appear more clearly in examples. A linear number is one beginning from two with unity added on. To one same and continuous line you add the accumulated quantities, that is, as we put below:

```
11    111    1111    11111    111111    1111111    11111111
             111111111                 1111111111
```

6. Concerning plane straight-line figures and that their principle is a triangle.

A plane surface is found in numbers[16] as often as beginning from three there is a width of a diagram created and numbers added on following each other in the multitude of natural numbers, and angles are extended, so that the first number is triangular and the second square, and the third is contained in five angles, which the Greeks call a pentagon, then a hexagon, that is a figure in which there are six angles. The remaining figures in this manner, one by one, through the natural numbers, increase their angles in outline of plane figures. These begin from the third number alone as the principle of width and surface. In geometry this same thing is found very clearly; two straight lines do not contain a space. Every figure is made of triangles, whether a tetragon or a pentagon, or hexagon, or whatever, which is contained by a number of angles, if from the middle to each angle are drawn lines until so many triangles divide the figure as that figure happens to have angles. In a square, lines divide the figure into four

---

16. For a definition of plane numbers, see also Euclid (*Elements*, Book 7, def. 17) who defines them as numbers which result from two terms multiplied with each other and which constitute the sides of each plane number. Theon gives a similar definition, p. 31. The definition of Euclid, in effect, excludes triangular and polygonal sides, and in this he differs from Nicomachus. Theon gives the definition of Euclid, but later lists polygonal numbers and so appears uncertain as what he means by the term. See D'ooge, p. 240 for a discussion. Phil Judaeus, p. 32, includes the wider definition, as Nicomachus. See a similar classification by Diophantus in *De Polygonis Numeris* (ed. Paulus Tannery, Leipzig, Teubner, 1893), Vol. 1, p.450. It is important to note that the same number may be considered as linear, plane, or solid, depending on how the monads are arranged.

triangles; in a pentagon, lines divide the figure into five triangles; lines divide a hexagon into six, and each other figure, in the mode of its angles and measure, is divided into triangles. So, in the following diagram, a square is divided into four triangles:

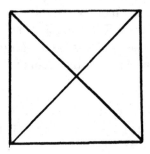

A pentagon is divided into five triangles

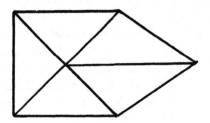

A hexagon is divided into six triangles

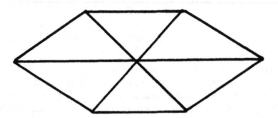

When someone divides a triangle, however, it is not resolved into other figures, only into itself. A triangle breaks down into three and is subdivided into three other triangles:

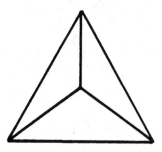

And so this figure is a principle of width and all the other plane figures are resolved into it; since it is obliged to no other principle nor takes its beginning from any other width, it is resolved only into itself.[17] The same happens with number, as the following part of my work will show.

7. The disposition of numbers of triangles.

And so there is the first triangular number, which is disposed in just three unities according to the description of the plane, that is with a triangular description; after this come whatever numbers separate equal sides in a triple spacing of sides:

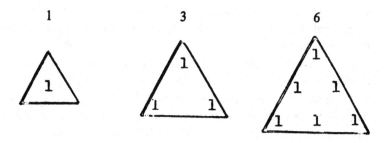

---

17. On the triangle as the basis of all plane figures, see Plato, *Timaeus*, Sec. 53. Plato goes further in saying that all triangles may be reduced to the elementary triangular form, the right-angle scalene or right-angled isosceles.

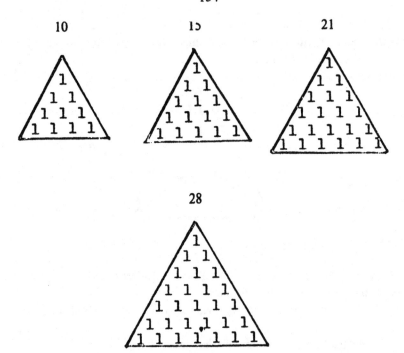

8. Concerning the side of triangular numbers.

According to this mode, there is an infinite progression and all the equilateral triangles are created in order. The first triangle which is born from unity, as here, is a triangle in power,[18] but not in act and operation. This power is, as it were, the mother of all numbers; whatever else occurs in the following numbers comes from that unity and it is found to be of those numbers. It is necessary that unity contains those numbers by a certain natural potency. The number three, which is the first triangle in operation and act, has the binary number as a side, in growing by a unity. Unity is the side of the first triangle, one in power and force; duality, which the Greeks call dyad, is the side of the first triangle in act and operation. The second triangle, that is the second in operation and act, the triangle of six, in growing by a natural number, has three in its sides. The third contains the side of four, that is the triangle of ten strokes. The fourth contains the side of five, that is the triangle of 15 strokes, and the fifth contains the side of six, and so on *ad infinitum.*

18. Nicomachus, Book 2, chap. 8, Theon, pp. 55-57

9. Concerning the generation of triangular numbers.

Triangles are born in the naturally disposed quantity of numbers if the multitude of sequential numbers is always grouped with its priors. The natural numbers may be disposed in this manner: 1 2 3 4 5 6 7 8 9. If I take from these the first term, that is unity, I have the first triangle, and that is a triangle in force and power, but not in act and operation. Now, if to this I join the second term which is described in the natural disposition of numbers, that is two, the first triangle is born for me, in operation and act, and that is the ternary triangle. If to this I add three from the natural number, the second triangle in operation and act is produced for me. If on top of one and two, I add the third, the number six is extended, that is the second triangle. If on this I impose the following four, ten is defined which is in act the third triangle, which you will note by the disposing of the sides in the exemplar of the above description, and so produce all the triangular numbers without fear of any doubt. However many unities the final number has in itself which you add to the previous, so many unities the triangle has, and whatever number the triangle is in order, so many unities will it have in its side.

Now three, which is the first triangle in act, we would make by the two added to unity, and this triangle would have two in its sides. We produce six with the quantity of three added to these, whose sides three alone contains. The same occurs in all the others; however many unities the number has, you join these to the previous triangular number, and by so many unities are its sides contained.

10. Concerning squared numbers.

A squared number is one which extends into a width but not with three angles as does the above figure. This figure is measured with four angles, in an equal dimension of sides. It is of this sort:[19]

19. Nicomachus, Book 2, chap. 9. For descriptions of squared numbers, see Theon, pp. 47-49; for their generation, see pp. 57-61.

11. Concerning their sides.

In these figures, the increases grow according to a natural number of the sides. The first figure is in power and potency a square, that is unity; it contains one on a side. The second, which is first in act, has four; it is contained by two on a side. All the others proceed according to the same sequence.

12. Concerning the generation of squared numbers and again concerning their sides.

Such numbers are born from the disposition of the natural number, not as the above triangles which were put together from numbers ordered to each other, but with one between always left out; if each is combined with the one before or the one after, of themselves they bring about ordered square figures. The natural number may be thus extended:

    1   2   3   4   5   6   7   8   9   10   11

If I look to the number one in this series, the first square number in power is born for me. Now if I join three to one with the number between left out, the second squared number emerges for me. So if with two left out and in a similar way I join five to four, the third of the squared numbers is created for me, and that is nine. For one and 3 and 5 give nine. Now if to this, with six left out, I add seven, the whole sum grows to 16, and that is the number of the fourth squared term. And briefly, the method of this procreation will be obvious if all the odd numbers are put together, with the natural order of numbers laid out; so the order of the squared numbers is produced. There is in this matter the subtle and immutable ordering of nature, that each of the squared numbers will retain so many unities in a side as there were numbers brought together for its creation. For the first squared number, since it came of one, has one in the side. In the second, that is 4, since it is created from one and three, which are two terms, each side is contained by two. And the same will be seen in the others.

13.Concerning pentagons and their sides.

That number is a pentagon which extends into width and is contained by five angles, described according to unity.[20] All its sides are disposed with equal dimensions. Such numbers are 1, 5, 12, 22, 35, 51, 70. Their sides grow in the same manner. In potency of the first pentagon there is one, and the same one maintains the space of a side. Of the second penta-

---

20. Nicomachus, Book 2, chap. 10; Theon, p. 67.

gon, five, which is a pentagon in act as well as in operation, there are two for each side. The third pentagon, which is 12, is increased by three for each side. The fourth, 22, is extended by the quantity of four numbers in a side. The same occurs in the remaining according to the progression of unity. Thus in the natural number they extend according to the increases of the above figures.

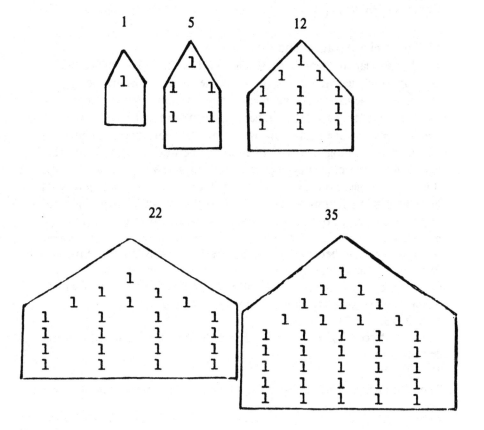

14. Concerning the generation of pentagons.

These are numbers which, extended in width, reach into five angles[21], and are born from that same quantity of the natural number added on to itself so that, with two left between each, the neighboring number is added to one or more previous numbers. If to the number one, with the numbers

---

21. Nicomachus, Book 2, chap. 10.

two and three left out, you add 4, which goes beyond unity by 3, the pentagon of five will be created. If after four, with five and six left out, you add seven, then you create the pentagon of twelve, for one and 4 and 7 give the number 12. This also happens in the other pentagon numbers. If you add 10, or 13 or 16 or 19 or 22 or 25 to all the previous in the same way as before, the result will be pentagonal numbers according to this order:
22, 35, 51, 70, 92, 117.

15. Concerning hexagons and their generation.

Hexagonal numbers are those contained by six angles and heptagonal by 7, and the augmentation of their sides develops in this manner. In the nature and procreation of triangular number we join together those numbers which follow each other in natural disposition and increase only by one. The creation of a square, that is a tetragonal figure, comes by numbers which are joined together with one left out between each, starting from two. The nature of pentagon comes from numbers beginning from the third and proceeding with two numbers left between. According to such increases also are figures of hexagons, heptagons or octagons or nine sided or ten sided figures or whatever other. These are all contrived from the progressive compilation of numbers. As in the pentagon, we join together numbers with two between each, and these increase by three's. Now in a hexagon we join them together with three between, and they surpass each other by four, and these will be the roots and foundations from which, when they are joined together, all hexagons are born:
1 5 9 13   17 21
and so on, according to that order. From these numbers are figures of six angles born:
1 6 15 28 45 66,
These you will find according to the above method, described in their proper order.

16. Concerning heptagons and their generation; the common rule of discovering the generation of all figures, and the descriptions of the figures.

A figure of seven angles is one which, when to the same order of progression with one more than the number left out between each in the figure of six angles, you join each number to the preceding. Now if with four between, beginning after the number five, you join each together, thereupon is the figure of the heptagon born, since these numbers are its roots and foundations, as we said before:

1 6 11 16 21.

And the heptagonal numbers which come from these are:

1 7 18 34 55.

The figures of nine angles are created according to the same order so that their numbers stand distant according to an equal progression of one. In the triangle, which is the first plane figure, there are numbers which precede each other only by one, that is they proceed according to their nature and description. In the tetragon, which is second, the numbers beginning after two are joined together, in the pentagon three, in the hexagon 4, and the heptagon 5 and in this matter there is no change. The diagrams of the following forms will show us this:

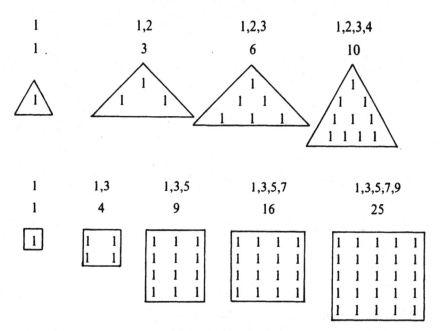

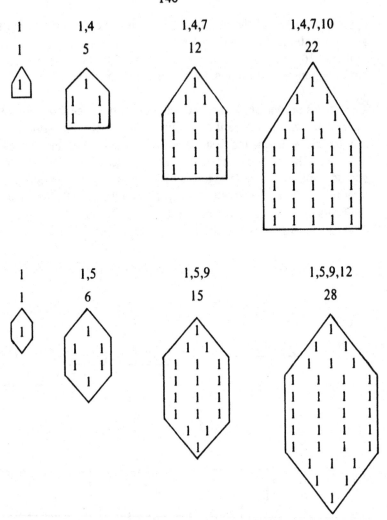

17. A description of the figured numbers in order.

Similarly we should be able to write out the quantities of the other forms which are contained by several angles.[22] In order that these data may be more easily grasped at sight, the numbers of these forms will be put in the following diagram:

22. Nicomachus Book 2, chap. 12.

Triang:	1	3	6	10	15	21	28	36	45	55
Quad:	1	4	9	16	25	36	49	59	81	100
Pent:	1	5	12	22	35	51	70	92	117	145
Hex:	1	6	15	28	45	66	92	120	153	190
Hept:	1	7	18	34	55	81	112	148	189	235

18. What figured numbers come from what figured number: that the triangular number is the principle of all the other numbers.

Since these things are so, we shall now investigate what follows from this matter. All tetragons which are disposed under triangles in natural order are created from the preceding triangles and the figure of the square is constructed from their assembly. The tetragon four comes from one and three, that is from the two superior triangles; nine comes from three and six, both triangles; 16 comes from ten and six; 25 comes from 10 and 15. The same is found in the continuing order of squared numbers and will be discovered to be constant and unchanging. The sums of pentagonal numbers are put together from a tetragon of four and from one in the order of triangles. The pentagon five is put together from a tetragon of four and from one in the order of triangles. The pentagon of 12 is born from a squared number nine and in addition to it three, the second triangle. The pentagon 22 comes from 16 and 6, that is a square and a triangle. 35 comes from 25 and 10. And to this order in the same manner of understanding, no hindrance of contradiction will impede the process. If you look at hexagons with a clear examination, you will see that they are created from the same triangles and pentagons added to them. The hexagon 6 comes from the pentagon five and one, which is a figure placed in the order of triangles. Nor is the origin of the hexagon 15 any different except that it comes from the pentagon twelve and the triangle three. If you understand that the hexagon 28 is born of certain previous figures you will find these to be none other than the pentagon 22 and the triangle six, and you will find this in the other numbers. Nor will the procreation of heptagon numbers deny this order of generation. Heptagons are created from hexagons and triangles imposed on each other. The heptagon number seven is born from the hexagon six and from the triangle in the power of one. The heptagon 18 is put together from the hexagon 15 and the triangle three; the heptagon 34 is created from the hexagon 28 and the triangle 6. This will be found without change in the other hexagonal numbers. So you will see that the first triangle produces the sums of all the figures and is involved in the procreation of them all.

19. A speculation pertaining to the description of figured numbers.

If all these numbers were compared with respect to numbers of sides, that is the triangles to the tetragons, or the tetragons to the pentagons, or the pentagons to the hexagons or these again to the heptagons, without any doubt each will exceed the others in terms of triangles.[23] What is the difference between the triangle three compared to four or the quadrangle four compared to five or the pentangle five compared to six, or six compared to the heptagon seven, except that they differ by the triangle one, that is, they exceed each other only by a unity. If six is compared to nine or 9 to 12 or 12 is compared to 15, or fifteen is compared to 18 -- if these are compared for their differences, they will exceed each other by the second triangle, that is by three. 10 compared to 16, and 16 compared to 22, and 22 to 28 and 28 to 34: if you put these together, they exceed each other by the third triangle, that is by six. And this will be fittingly noted in all the following numbers and each figure will precede the other in terms of triangles. Thus it is perfectly demonstrated, as I would judge, that the triangle is the principle and the basic element of all forms.

20. Concerning solid numbers.

From this point, the way to solid figures is easier.[24] You should know that the force of quantity which operated naturally in plane figures of numbers will apply without change to solid numbers. Just as there is another dimension to the length of numbers, that is the plane, as its width demonstrates, so now if someone adds to width that dimension which some call altitude, others call thickness, others call depth, it will define the solid body of numbers.

21. Concerning the pyramid and that it is the principle of solid figures as the triangle is of plane numbers.

It will be seen that just as in plane figures where the triangle is the first number, so in solids, that figure called the pyramid is the principle of depth. It is necessary to find the primordial element of all three-dimensional figures in numbers. Such is a pyramid, that is, a figure based on a triangle and rising to a height; it is the basis of a tetragon or pentagon and

---

23. Boethius' text reads: *si ad latitudinem fuerint comparati*. *Laditudo* is usually width, but here the explication makes clear that Boethius is speaking of the number of sides. It seems that *ad latitudinem* designates the category of plane numbers.
24. Nicomachus, Book 2, chap. 13-14

of a multitude of other figures according to the sequence of angles, raised to one point of height. With a triangle written and described, if we place single lines from the three angles, first standing straight then inclined so that they join at a vertex on one point in the middle, a pyramid is produced. Thus is it constructed from a triangular base, and is closed in by three triangles on the sides in this manner: ABC represents a triangle. On this triangle lines are erected through the three angles and are converged toward one point, that is D, in a way that the point D is not on the plane but elevated from it. The lines straight to the vertex create a sort of peak at D so that the base ABC is one triangle, and there are three triangles along the sides: one is ABD, another is BDC, and a third is CDA.

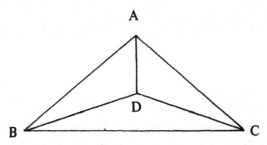

22. Concerning those pyramids which are made from squares or other multiple angled figures.

If the same lines are led from a tetragonal base and the lines are directed to one vertex, there will then be a pyramid of four triangles on the sides and one tetragon placed at the base, upon which the figure itself is established. If from a pentangle there arise five lines, the pyramid will be contained by five triangles, and if it rises from a hexagon there will be six triangles. However many angles a figure has on which a pyramid rests, by that many triangles is it contained on its sides. This is clear in the following descriptions.

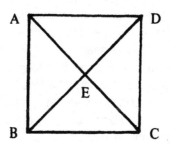

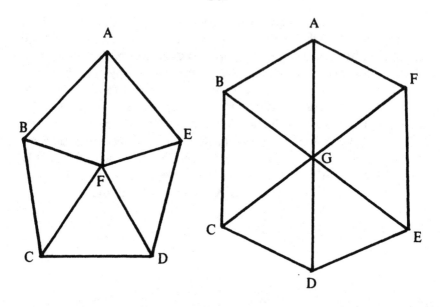

23. The generation of solid numbers.

Pyramids of this type are named in the following way: the first is a pyramid from the triangle, the second a pyramid from the tetragon, the third pyramid comes from the pentagon, the fourth from the hexagon, the fifth from a heptagon, and the same remains constant in the remaining numbers. Since we said that numbers are linear and, drawn out from one, run to infinity, such as are 1, 2, 3, 4, 5, 6, 7, 8, 9, 10, it is from these numbers put in order that the plane figures are born when they are joined to each other with distances between each. So if you join one to two, the first triangle is born, that is three; to these numbers we would join the third, and the triangular number of six occurs, and after these, with one skipped over, the tetragons are born; with two skipped over, the pentagons are born; with three, the hexagons are born; with four, the heptagons. Now, for the creation of solid bodies the figured numbers naturally provide surfaces for us. In making pyramids from a triangle, we must put together those same triangles. For making pyramids from tetragons, tetragons must be put together; for pentagons, pentagons must be put together. And those figures which come from hexagons and heptagons are born only from hexagons and heptagons put together. The first in power is the triangle and unity, so in terms of its power we put the pyramid in the same place. If I join the second triangle, the triangle three, to the first which is unity, the four-fold depth of the pyramid will emerge for me. If to this I add the third triangular number, which is six, the depth of the pyramidal number ten will be created. If to these I add ten, the pyramid of 20 numbers will emerge, and so in all the others there is to be found this ratio of joining numbers.

Triangles:

| 1 | 3 | 6 | 10 | 15 | 21 | 28 | 36 | 45 | 55 |

Pyramids coming from these triangles are:

| 1 | 4 | 10 | 20 | 35 | 56 | 84 | 120 | 165 | 220 |

In this kind of joining it is necessary, as always, that the number which is last of those joined together exists as a basis. It will be found wider than all the previous. And it is necessary that the numbers joined before it be smaller, all the way until the reducing proportion comes to unity, which maintains the place of a point or vertex. In the pyramid 10, to six are added three and one, which ten exceeds six first by the quantity of three, then unity offers the final term of the progression. A similar reasoning is able to be seen in the rest, if you would be willing to scrutinize their procreation diligently. These pyramids are born by the same composition of tetragons upon themselves. Let us describe all these pentagons:

| 1 | 4 | 9 | 16 | 25 | 36 | 49 | 64 | 81 | 100 |

Now let us look at unity in this disposition; the first pyramid produced for us in power and force is unity, but it is not yet in act and operation.[25] Now if to this I add the tetragon, that is 4, the pyramid of five numbers results, which is contained by just two numbers on the sides. If to these I add the following nine, the result will be the form of a pyramid of 14 numbers which is enclosed on the sides by three unities. If to this I add the following tetragon, I will produce 16, and a three sided form of a pyramid emerges. In all these pyramids there will be so many unities per side as there are quantities of numbers added together in it. Now unity, which is the first element of a pyramid, is that which presents itself as one side; five, which is composed of one and four, is designated by two sides and 14 which is composed of three numbers, is established with three sides. The following description shows the generation of the pyramids.

Tetragons:

| 1 | 4 | 9 | 16 | 25 | 36 | 49 | 64 | 81 | 100 |

---

25. This discussion of pyramids is considerably expanded by Boethius from his source.

Pyramids from the tetragons:

1	5	14	30	55	91	140	204	285	385

According to this same manner, all the increasing forms from the remaining multi-angular figures are directed into spaces of a higher number. Every form of a multi-angular number from a figure of its own type, with unity imposed by increases from one, proceeds into infinity in assembled figures of pyramids. From this it is necessarily obvious that triangular forms are the principle of other figures because every pyramid, from whatever base it is constructed, whether from a quadrangle, pentagon, a hexagon, a heptagon or whatever similar figure, is contained only by triangles all the way to its vertex.

24. Concerning truncated pyramids.

It is important to know that truncated pyramids are, as well as twice-truncated, three times truncated, four times truncated, and so on according to the addition of numbers. A perfect pyramid is one which, arising from a given base, reaches to the first pyramid in force and power, unity. But if the altitude does not come from a given base to unity it is called truncated and a pyramid of this kind is rightly designated by such a nomenclature if it does not reach to a point and extremity. Such a form would exist if someone would add to a tetragon 16, and 9 to 4, since these hold themselves back from joining with the ultimate point of unity. It is the figure of a pyramid, but since it does not come to a vertex, it is called truncated and it will have as its top not a point, which is unity, but a plane which is a certain number extended and put together according to the angles in the base of the triangle. So if a tetragon were the basis, the square would rise gradually by diminution, and if a pentagon is the base, similarly, and if it were a hexagon, so would the top plane be a hexagon. Therefore in a truncated pyramid, the plane will be of so many angles as there were in the base. If that pyramid does not only fail to come to its extremity and the point of unity, but does not even come to the first multiple angle of its origin, of which it takes its name from the base, it will be called twice truncated. If starting from a tetragon of 16 it makes an end at nine, and does not go to four, then by however many tetragons it is short, by so many do we call it short, as the Greeks say κολοῦρον [truncated]. If it is lacking two tetragons, that is unity and another one, then that which follows is called twice truncated, which the Greeks call δικολοῦρον . If it is short by three tetragons, it is called thrice truncated, which the Greeks say is τρικολοῦρον . However many tetragons it is lacking by so many is that pyramid said to be truncated. This is true not only in a tetragonal pyramid but it is seen in all pyramids proceeding from every multiple angled figure.

25. Concerning cubic numbers, »beams,« »bricks,« »wedges,« spherical numbers, and parallelepipedic number.

Concerning solids which take their shape from pyramids, it is said that they proceed equally from their own multiangular figure as from a root.[26] There is a certain other ordered composition of solid bodies, of those which are called cubes, »beams,« or »bricks,« or »wedges,« or spheres or parallelepipedic numbers which, since their surfaces are separated from each other when led to infinity, never fall together.

Put into order here are the tetragons 1   4   9   16   25, and these are distinguished because they take on only width and length, and lack depth. If they should receive only one multiplication on the sides, they would proceed to an equal depth. The tetragon four has two in its side and is born from twice two; twice two makes four. If you multiply these two from the side of the figure equally, the form of the cube would be born. If you make twice two times two, the quantity of eight grows from it, and this is the first cube. The tetragon nine has three on a side and is made from three multiplied by itself; if you add to it one multiplication of the side, immediately there will issue another cube with an equal formation of sides. Three times three, done three times, and the cubic figure 27 is produced. The tetragon 16, which comes from four, if increased four times, rises to the cubic figure 64, with equal measure of its sides. The following tetragons will be produced through multiplications made in the same manner. It is necessary that a cube have as many unities in a side as the first tetragon from which it was produced had in its sides. Since the tetragon four has only two numbers in a side, the cube of eight has also two. And since the tetragon of 9 is marked on its side by three unities, then so many unities does the cube of 27 present in its side. Since a tetragon of 16 has four unities in its side, the cube of 64 generates that same number in its side. In a unity which is the power and force of a cube, only one will be in a side. Every tetragon is a single plane of four angles and as many sides. Every cube which grows into depth from the plane of a tetragon, multi-

---

26. These terms are taken from D'ooge's translation of Nicomachus. Hero of Alexandria uses the term "beam" (δοκίς) and defines the figure (def. 112) as a solid having a length greater than the breadth or thickness and the latter two are sometimes equal. Theon uses the same term with that meaning. The Greek terms σφηνίσκος, σκαληνός or σφηκίσκος correspond to "cuneus," and "sphensicos," all used by Boethius to mean types of figures longer by one side. In all these cases, Boethius provides ample definitions to clarify his meaning.

plied through the side of a tetragon, has six planes, of which each single surface is equal to the prior tetragon and has 12 sides, of which each single one is equal to those of the original tetragon. As we have shown above, it also has eight angles and each of these is contained under one of the three types to which the prior belonged in the original tetragon from which the cube was produced. Thus, tetragons are provided in the following diagram from the natural extension of number and from these tetragons come the cubes which are noted below.

Natural numbers:

1	2	3	4	5	6	7

Tetragons:

1	4	9	16	25	36	49

Cubes:

1	8	27	64	125	216	343

Since every cube is put together from equilateral quadrangles,[27] each is of itself equal in all its parts, for it is equal in width, length, and height; and it is also equal according to its six sides, that is, below, above, right, left, front, and back, of necessity. Opposed and contrary to this sometimes it would be that a figure has a length that is not equal to its width, or two sides that do not extend equal to the depth, but all would be unequal and such a figure would be considered to be a long way from the equality of a cube. Such a figure would come about if one should make twice three times four, or three four times five, and other multiplications of this sort, which are unequally produced through unequal steps of spaces. These forms are called by their Greek name »scalene.« We may call them graduated forms because they grow, as it were, by steps or grades from smaller to greater. The Greeks call this same figure a »spheniscon;« we may also call it a »cuneum.« They shape such forms in constructing a thing by no regular ratio of width, length, or height; but as much as it is fitting, so much is given to the height, or so much is it increased in extent of depth. It is necessary and important to find these forms unequal in every part. Some people call them »bomiscoi,«[28] that is as some sort of small altar which in the Ionic region of Greece, as Nicomachus says, was shaped in

---

27. Nicomachus, Book 2, chap. 16; Theon, p. 69.
28. Nicomachus has βωμίσκος or "little altar," a scalene figure having both sides and angles uneven. Boethius also uses the term "bomiscus" and "arula" with this meaning. See also Theon, p. 71. Heath (Elements, Vol. 2, p. 290) disucsses the term as it appears in Euclid.

such a manner that neither height, width, nor length were the same. They were also called by certain other names which we think it superfluous to pursue now. Therefore cubes, constructed with spaces equal to each other, and these forms, which we have spoken of, are in a middle position, disposed with a graded distribution, which are not equal in all parts, nor unequal in all parts, which the Greeks call »parallelepipedons.« The Latins do not have a name for them that is uniformly accepted, so that by many the figure is called by the Greek term. The figure called by this name is contained by irregularly placed sides.

26. Concerning numbers longer by one side and their generation.

Forms of this sort are the kind which the Greeks call ἐτερομήκεις, we are able to call »longer by one side.«[29] The number of such a figure must be defined in this manner: you have described a number longer by one side if it has width and it has four sides and four angles, but these are not equal and they are always one smaller than the other. Nor are all sides equal to all sides, and latitude is not equal to longitude, but as we said, one side is greater and surpasses or supercedes the smaller by one measure. If we put the natural numbers in order and multiply them by the first, such a number will emerge; if you multiply the second by the third, the third by the fourth, the fourth by the fifth, and all with just unity added, then are born the numbers larger by one side. So let the natural numbers be thus disposed:

1	2	3	4	5	6	7

Now let us proceed. If someone multiplies one by two, he makes a 2; if he multiplies two by three, he makes 6; three times four makes 12; four times five makes 20; and so on, according to the same order. Whatever numbers are thus made are created with one side longer, as the following diagram shows; in it numbers longer by one side emerge when those are multiplied which are written above, and those are born which are written below:[30]

29. Nicomachus, Book 2, chap. 17; Jordanus, Book 7, prop. 17; Book 9, prop. 38. Boethius uses the term *numeri altera parte longior.* See the diagram from Book 1, chap. 26, above, where the number 9 is a square and there 3 meets with 3; at 12, a number (or figure) longer by one side, 3 meets with 4. such figures may also be represented as n (n + l).

30. For the generation of numbers longer by one side see also Theon, p. 49.

1	2	3	4	5	6
2	4	6	8	10	12
3	6	9	12	15	18
4	8	12	16	20	24
5	10	15	20	25	30

1	2	3	4	5	6	7
2	6	12	20	30	42	
1 1	1 1 1	1 1 1 1	1 1 1 1 1	1 1 1 1 1 1	1 1 1 1 1 1 1	
	1 1 1	1 1 1 1	1 1 1 1 1	1 1 1 1 1 1	1 1 1 1 1 1 1	
		1 1 1 1	1 1 1 1 1	1 1 1 1 1	1 1 1 1 1 1 1	
			1 1 1 1 1	1 1 1 1 1 1	1 1 1 1 1 1 1	
				1 1 1 1 1 1	1 1 1 1 1 1 1	
					1 1 1 1 1 1 1	

27. Concerning »antelongior« numbers and the terminology of the number longer by one side.

So, if numbers differ by only a unity, when they are multiplied together, the numbers described above are created, unless you multiply by some other number, as three times 7 or three times 5 or in some other manner, and their sides do not differ by only a unity. Then they are not called by the term longer by one side, but »antelongior.«

By Pythagoras, or by the heirs of his wisdom, they are described with no other number than by two. They call this the principle of alternity and say that the same nature always follows identical to itself and becomes no other except through ungenerated primal unity. The binary number, the first real number, is different from unity because at first it is separated from it by a unity. And so this is the principle of alternity because it is dissimilar from unity, the first and always same substance. Rightly is it said therefore that these numbers are longer by one side because of the sides of such a figure which precede each other by a single adjacent numerical value. There is an argument, however, that alternity may justly be constituted in the binary number which is said by some persons to be only from two, and among these persons there is no lack of reason in such argument. But again, it has been demonstrated that the odd number alone is produced from unity, and that the even number is produced only by duality, that is by the binary number. The median of any number is one if it is an odd number; if the number is even, this equality is divided into equal halves. So it must be said that of the even numbers there is an odd number participating, and that if it participates in the immutable substance of its nature, then the even number is formed from unity and the even number is full of the nature of the other and because of that, it is completed in duality.

28. That from odd numbers, squares are made and from even numbers, figures longer by one side are made.

With the odd numbers put in order from unity and the even numbers written under these from two, the adding of odd numbers makes tetragons and the putting together of the even numbers below them brings about figures longer by one side. Since this is the nature of tetragons, that they are created from odd numbers which participate in unity and are of its immutable substance; they are equal in all their parts because angles to angles, sides to sides, and length to length are equal, it must be said the numbers participate of this same nature and substance. Those numbers which equality makes longer by one part we may say are of another substance for just as one from two is different by one, so the sides of these figures are one more than the other and differ only by a unity. So here are disposed in order odd numbers from one and under these all even numbers beginning from two.

1	3	5	7	9	11	13
2	4	6	8	10	12	14

Unity is the principle of the odd order because it is itself a certain effective cause and as it were a certain form of the odd number insofar as it is of one and immutable substance. So, if it is multiplied by itself or on a plane or in the height of a figure or it is mulitplied by some other figure, it does not differ from the form of its prior quantity. If you make one the same, or the same one once or two the same, or three the same, or four the same, or whatever other number it multiplies, that number does not differ from the quantity which it multiplies because it will not assume an even or odd status different from the number it multiplies.

But the even number is the principle of the binary order because duality, since it is in the same order of equality, is as well the principle of every alternity. If it multiplies itself by the width or even the depth, or it combines some number into its quantity, then that number immediately becomes different. For if you make twice one or twice two or twice three or twice four, or twice five, or multiply whatever other number, whatever is born in this way is found to be other than what it was at first. All tetragons are born in this way from the above diagram and from the first order. If you look at one, it is the first tetragon in power. If you put one together with three, the tetragon of four emerges. If to it I add five, nine will result. If to this I add seven, the form of sixteen is furnished. If you do the same in the rest of the numbers, you can see all the square numbers created in order.

From the second order of even number all the figures longer by one side emerge. For if I look at two in the first place, a number of this sort occurs for me which comes of twice one. When with two I join the following number four, there will immediately emerge a figured number longer by one side, namely six, which comes from twice three. If to this I add the following number, the form of 12 emerges for me, which comes from four times three. If someone should do this continually, he would see all the numbers of this sort created in proper order, and the following diagram demonstrates this description:

Roots:	1		1,3	1,3,5	1,3,5,7
Tetragons:	1		4	9	16

Roots:	1,3,5,7,9		1,3,5,7,9,11
Tetragons:	25		36

Roots:	2,4	2,4,6	2,4,6,8
Longer by one side:	6	12	20

Roots:	2,4,6,8,10		2,4,6,8,10,12
Longer by one side:	30		42

29. Concerning the generation of »latercular« numbers and their definition.

Above, we have called those numbers »latercular« which are of a solid figure but are made in this way: to spaces extended equally in width and length there is added a smaller altitude. Such figures are of this sort: three three's twice, which make 18; or 4 four's twice; or any other combination so that such figures are extended into length and width by equal distance, but into depth by a smaller distance. Such figures are defined in this manner: those are latercular which from equals are equally made smaller. The »asseres« are figures which are solid but in such a way that from equals they equally extend into a larger side. If a figure has a length equal to the side but there is a greater depth, we call such a figure »asseric« and the Greeks call it »docedic.« Let us proceed in this manner: 4 fours nine times—and the figure produced by this process is called »asseric.« »Spheniscal« numbers, which above we have called »cuneol,« are those produced by sides not equal to length, extended to an odd depth, while cubic figures are produced from equal sides, equally extended to equal distances.

30. Concerning circular or spherical numbers.

When the numbers of cubes are so extended that from any number of cubic quantity a side begins, and the extremity is terminated at the same point of height, then that number is called cyclical or spherical. Such are the multiplications which begin from five or from six. Five times five, which make 25, having progressed from 5, ends at the same 5. If you extend this five out again, its terminus will again come to 5. Five times 25 makes 125 and if you bring this number to five times more, it will be terminated in a five number. And this will always happen the same way, all the way to infinity. It would also be suitable to consider this process in the number six.

These numbers are called cyclical or spherical because spheres and circles are always formed by a turning about of their own principal number. A circle is a figure which, with one point put down, and another placed at a distance from it, that placed at a distance is brought around, all along with an equal space from the middle point constantly maintained so that it ends at the same point again from which it began to move. A sphere is a figure which is produced by a semi-circle remaining on the same diameter and turned around until it returns to the same point from which it first moved.[31]

Unity in both power and force is a circle and a sphere. As often as you multiply a point by itself, it always ends in itself from which it began. If you multiply one by one, one remains; and if you multiply it again and again, it is still the same. If there is one multiplication of a number, it gives a plane figure, which is a circle; if you multiply it a second time, then a sphere is created. The second multiplication is always the producer of the depth in a figure. Therefore from 5 and from 6 we will write down short diagrams of this sort:

1	5	6
1	25	36
1	125	216
1	625	1296
1	3125	7776

31. This definition is not found in Nicomachus.

31. Concerning the nature of those things which are said to be of the same nature, and concerning the nature of things which have different natures and are joined to a number of the same nature.

The things said at present will suffice concerning solid figures.[32] Those thinkers who investigate the nature of things with close reasoning and who are versed in mathematical disputation have instructed us in a most subtle and learned way about what may be considered the property of any one thing. These people distinguish the natures of all things, dividing them by this reasoning into two. They say that all substances of all things consist in that which is always of its own proper habit, and this can in no way be changed, and such is a nature which fixes the substance of variable movement. They call this the first immutable nature of one and the same substance. Another which departs from that first immutable substance is a second nature which still truly pertains to unity as well as to duality which is the first number departing from one and made another. Since all odd numbers were formed according to the nature and type of unity, these tetragons which are developed from this kind of formation are said to be participators of this substance in a double way: either the tetragons are formed from equality or they are created from odd numbers added together into one figure. Those tetragons which are even, since they were formed of the binary number, and those which were collected and gathered together into one accumulation, come into being as numbers longer by one side. These are said to be separated according to the nature of that same binary number and from the nature of the same substance, but they are considered to be participants of another nature, because the sides of a tetragon progress from equality into equality by virtue of their own width. Figures longer by one side with one number added on are not equal in their sides but are constructed with dissimilar sides and, as it were, joined together with something other than themselves. Thus it is known to us, that just as it is in this matter, so in the world are things joined together. Either things are of the same immutable proper substance, as are God, the soul or the mind, or whatever is blessed with incorporality by its own nature, or they are of a variable and mutable nature, which we undoubtedly see is the case in corporeal things.

So we must show what this double nature of number is, that is, what are squares or any other type of figure longer by one side, or what rela-

---

32. Nicomachus, Book 2, chap. 18.

tionships they may have when they are considered in regard to each other. We shall see whether they are multiplex, superparticular, or whatever. When considered in themselves, as the forms are shaped in the way we described above a little while ago, we shall see whether they are from a mutable or an immutable substance. So, every number from the tetagrons, which are made of unchangeablness, and the oblong quadrangles, which participate in mutability--it is relevant that these things be examined. First, the figures called »promeces« must be sorted out; these are the figures longer by the rear part. Then there are those called $\dot{\epsilon}\tau\epsilon\rho o\mu\dot{\eta}\kappa\epsilon\iota\varsigma$ that is, longer by one side. There is a figure longer by one side whose sides grow by a unity added on, as are six, that is twice three, or 12, three fours, and similar numbers. The figure longer by the rear is one which is contained by two numbers of this kind and the differences between certain other numbers, as three fives, or three sixes or four sevens. It is as though the number longer by an anterior part is said to be extended in a multiple mode. We have said above why certain numbers are called longer by one side. Since squared numbers have equal length and width they are most aptly called by their length or width, as twice two, thrice three, four fours, and so on. Those longer by one side, since they are not extended by an equal length, are called longer by one side as though composed of a different length.

32. That all things consist of the same nature and then of the nature of another, and that this can first be seen in numbers.

A certain thing which, though it be immobile in its own nature and substance, is terminated and defined even if it is not changed by any variation, cannot cease to be, cannot ever be, because it never had existence. This is only unity, which is formed by unity, and is said to be of a comprehensible, determined and constant substance. There are those substances which grow from equal numbers, such as squares, or those which unity forms, that is the odd numbers. But the binary numbers and figures longer by one side which are separated from the finite substance are named in virtue of a variable and infinite substance. So it is that every number which comes from these figures, which are disjunct and contrary comes, from even and odd numbers. One is a stable and the other is an unstable variation; the first has the strength of an immobile substance, the other is a changing mobility. The one is a limited solid, the other an infinite aggregation of multitude. These things which are contrary yet are mixed into a sort of comradship and friendship; by the formation and putting together of their unity, they produce one body of number.

Therefore not uselessly nor improvidently are similar natures from this world and this common nature of things so linked in my reasoning. These numbers have first determind this division of the substance of the world. In the *Timaeus* [33] Plato discussed things of one nature and of their opposite. In this world he thinks some things endure by virtue of their own nature as undivided and unconnected and as in the first element of all things; other things he considers as divisible and never remaining in the state of their own order.

Philolaus[34] said that all things necessarily are infinite or finite when he wished to demonstrate that all things, whatever they are, are made up of two things, either of an infinite or of a finite nature, without any regard to a similitude of number. These figures are joined from the numbers one and two, from odd and even, which are manifestly of equality and unequality, and so of one and the other, they are substances limited and unlimited. Thus not without reason is it said that all things which consist of contraries are compounded and joined together by a certain harmony. This harmony is the union of many things and the consensus of dissidents.[35]

33. From the nature of the same or of another number, in what do the relationships of a figure longer by one side and all the habits of proportion in it consist.

Let there be disposed in order not yet the even and odd numbers from which are produced squares or figures longer by one side, but those which when brought together are reduced to one.[36] It is from them that squares and figures longer by one side come. We thus will see a certain consensus of these numbers and that they produce a friendship toward other kinds of numbers, so that in this matter, not without reason, can the nature of things be seen as if incorporated in a type of number. Therefore let two rows be put down, one of the tetragon order from unity, then a row of numbers longer by one side:

1	4	9	16	25	36	49
2	6	12	20	30	42	56

33. *Timaeus*, Sec. 35A.
34. A Greek Pythagorean of the fifth century B.C.
35. On The notion of blending opposites or contraries in the creation of the universe, see Plato, *Laws*, Sec. 889B, where he describes the fusions of hot and cold, dry and moist, soft and hard, etc. For a discussion of contraries in behavior as well, see *Protagoras*, Sec. 332A-B. Opposites in the notion of the beautiful are discussed in *Republic*, Sec. 475-76.
36. Nicomachus, Book 2, chap. 19.

If you compare the first to the first of these, the quantity of duplex will be found, which is the first type of the multiplex; if the second is compared to the second, a hemiola relationship is produced. If the third is compared to the third, a sesquiterital ratio is created; if the fourth to the fourth, a sesquiquartan; if the fifth to the fifth, a sesquiquintal. You will find this rule of the superparticulars proceeding, to whatever proportional distance, whole and unchanged. So in the first duplex ratio there is a difference of only a unity. In the sesquialter relationship there is a difference of two numbers. In the sesquitertian there is a difference of three, in the sesquiquartan of four, and so on according to the superparticular forms of these numbers, which adhere to fixed differences. The addition of numbers grows by one only, outlining the natural order:

	1		4		9		16		25
1		2		3		4		5	
	2		6		12		20		30

If you compare the second tetragonal number to the first figure longer by one side and the third to the second and the fourth to the third and the fifth to the fourth you will note that those same ratios are produced which we described in the diagram above, yet here the differences do not start from unity but from the binary number and proceed by the same calculation into infinity. The second to the first will be in a duplex relation; the third to the second in a sesquialter; the fourth to the third, sesquitertian according to the same pattern that was shown above.

	4		9		16		25		36
2		3		4		5		6	
	2		6		12		20		30

Again the squared numbers differ between themselves by uneven proportions, the longer by one side differ by even.

Odd differences

	3		5		7		9		11		13	
1		4		9		16		25		36		49

Even differences

	4		6		8		10		12		14	
2		6		12		20		30		42		56

Longer by one side

Now if between the first and second tetragon we put the first figure longer by one side, it is joined to both of those by one proportion. In each of these proportions, the multiplex of a double is found. If between the second and third tetragon you put the second figure longer by one side, the form of the sesquialter comparison is assembled. If between the third and fourth tetragon you set the third figure longer by one side, an example of the sesquitertian is established. If you do the same in all the numbers, you will marvel to find all the superparticular types.

	Duplex	
1	2	4
	Sesquialter	
4	6	9
	Sesquitertial	
9	12	16
	Sesquiquartus	
16	20	25

This can be seen in the other numbers according to the same manner. Again if two tetragons are put down from the above diagram, that is the first and second, and they are collected together, their middle term, a number longer by one side, is multiplied twice and makes a tetragon. If one and 4 are joined together, they make 5. Of those numbers, two is a figure longer by one side; if it is multiplied by two it becomes 4. If joined to the previous number it makes 9, without fail, and this is a square number. This continuous process is understood in the other numbers according to the same manner, with terms disposed in the order which we have described above.

Between the two, exchange the first and second, and place the tetragon second to the figure longer by one side, which is second in order, but in act and operation is first. Then from two figures longer by one side, put together and multiplied by two with a tetragon between, immediately a tetragon is made. Between the numbers six and two which are the first and second of the numbers longer by one side, if the tetragon is second in the order, first in act, and 2 and 6 are added, they make 8; if then the middle number 4 is multiplied by two, it makes 8; and if that is joined with the previous number longer by one side |8| they elaborate the tetragon 16.

	5			13			25	
	4			12			24	
1	2	4	4	6	9	9	12	16
	9			25			49	

	8			18			32	
	8			18			32	
2	4	6	6	9	12	12	16	20
	16			36			64	

It is important to note with no less admiration that according to their own proper natures, where two tetragons stand opposite each other and a figured number longer by one side is placed in the middle, the tetragon which is born from this process is always created from odd numbers. From the above one and four and the two multiplied twice, the tetragon of nine is made which is created, as it were, from three. Three times three makes 9 which is the third odd number. Then the number which from 4 and 9 is joined with six multiplied by two gives the tetragon 25 and is itself born from the odd number 5; that number is the next after three. Five times five produces 25 and five is the next odd number after three. The logic for the following numbers is the same. The number produced from 9 and 16 and two multiplied by 12 is the square number 49 and that number is the seventh odd number, the next after 5. Seven times seven gives 49. Where two figures each longer by a side next to each other include a tetragon between, all the tetragons produced from these two figures are produced by even numbers.[37]

The tetragon which is made from the number longer by one side, from two, six, and four multiplied twice, is 16; and that is produced by the number four which is an even number. Four times 4 is sixteen. In the following set, where to 6 and 12 and their sum is added two times nine, the total is 36; this sum is put together from the even number six multiplied by itself. Six times six gives 36. The tetragon 64 falls no less into the same reasoning; it is made from 12 and 20 and twice 16; this is born from eight multiplied by itself, which is the first even number after six. Eight eights multiplied together give a tetragon of 64. If in the other numbers you proceed according to this manner, this order of reasoning will not change.

34. From squares and from figures longer by one side the idea of every form takes its being.

The fact that the entire development of all forms may be seen to arise from these two forms should be noted with no small consideration.[38]

---

37. In this chapter, as in most of the others in Book II, the number diagrams are not found in Nicomachus but have been added by Boethius.
38. Nicomachus, Book II, chap. 20.

Triangles variously put together produce all the other forms, as we demonstrated above; every form arises from triangles put together. Thus, from the 2 of the first figure longer by one side, the triangle of three is put together; from two and four, the second triangle of six is created. From four and six the triangle ten is born, and according to that order, the whole ratio of triangles stands. Let there be disposed alternately in two ranks between themselves tetragons and figures longer by one side so that this may be better noted. Then we will mix them and note down the triangles which arise in the process:

Tetragons

1	4	8	16	25	36	49	64	81

Figures longer by one side

2	6	12	20	30	42	56	72	90

Triangles

	3	6	10	15	21	28	36	45	55	66	78
1	2	4	6	9	12	16	20	25	30	36	42

35. How squares are made from figures longer by one side; how figures longer by one side come from squares.

Every square, if it has its own width added to it, or if some amount is removed from it, becomes a figure longer by one side. Now if to the tetragon four, a person adds two or takes two away, it becomes 6 by adding and two by subtracting. Each square has in it a figure longer by by one side, so effective is the power of change. Every infinite and indeterminate power differing from the nature of equality and from the substance contained in its boundaries either grows greater or becomes smaller.

36. That unity is the principal substance, odd numbers are in a second place to it, and the tetragon is in a third place; duality is another substance, even numbers are in second place to it, and figures longer by one side are third.

It happens therefore, in the first place, that unity is the principle of proper and immutable substance and nature, and duality is the first principle of difference and change. In the second place all odd numbers are participants on account of the knowledge of unity and its immutable substance; even numbers are mixed with their differences, in consort with binary numbers. It is obvious that tetragons are considered to be of the same mode. By virtue of the fact that their composition and conjunction comes from odd numbers, I will call them all of an immutable nature. Because figures longer by one side are created by the joining together of even figures, they are never separated by the variation of a difference.

37. What is the consensus of differences and proportions in squares and figures longer by one side when these are placed opposite each other.

It must be seen that if at the same time tetragons and figures longer by one side are disposed so that they are mixed alternately between themselves, such is the joining together in these that they communicate with each other in the same proportions. They are separated by the same differences but by other differences they are the same and these stand in certain proportions. Let the tetragons and the figures longer by a side be disposed in order from one:

| 1 | 2 | 4 | 6 | 9 | 12 | 16 | 20 | 25 | 30 |

In the above formula this especially must be seen: between one, which is a tetragon, and 2, there is duplex proportion; between 2 and 4 there is another duplex. The tetragon with one side longer is joined with the following tetragon in the same proportion but not by the same differences. Between one or two there is a difference of one, but between two and four there is a difference of two. Again if you look at two in relation to four, there is a duplex proportion, but between four and six you see a sesquialter proportion. Here they differ in proportions, but they are equal in their differences.

Four stands from two and six stands from four by the same two. In the following numbers by the same method, as it was in the first, this ratio holds firm. There is the same proportion, but not by the same differences. Four to six, 6 to 9, are joined by a sesquialter proportion, but six exceeds four by two, 9 goes beyond six by three. In the following numbers the same ratio will be seen always in an alternating fashion. There will be the same proportions but other differences; again with the order changed, it will be in the same differences but other proportions. In whatever they differ, the tetragons and the forms longer by one side will always exceed each other according to a progression of the sums of the natural number. No one should be surprised to see this. We have doubled the sums of the tetragons and figures of one side longer, to the first and second proportions:

Proportions

	duplex		duplex		sesquialter		sesquialter		sesquitertian
1		2		4		6		9		12
	1		2		2		3		3	

Differences

Proportions

sesquitertian		sesquiquartan		sesquiquartan		sesquiquinta		sesquiquinta	
12		16		20		25		30		36
	4		4		5		5		6	

Differences

These same differences proceed according to this marvelous manner from the whole through the following parts and through the same unities by which the above comes into existence. Between one and two only a unity interposes; of the unity, to which it is equal, it is whole: it is half to two. In the same way, between two and four there is 2, which is the entirety of the number two, half of four. Between four and 6, there is the same two, which is half of four and a third part of six. Between six and nine, which follows, there is three, and it is half of six and a third part of nine. Then three, which is a third of nine, comes next which is also a fourth part of twelve. According to this same manner you see the equal progression of parts to the end of the diagram, with double ranks of this sort, as they are created by the comparisons of sums.

38. A proof that squares are of the same nature.

This is a most evident sign that all tetragons are known by odd numbers, that in every disposition from one, either in doubles or triples in which such a natural order is established, never is the tretragon found except according to an odd order. Let us then put down numbers in order, first the doubles then the triples:

1	2	4	8	16	32	64	128	256
1	3	9	27	81	243	729	2187	6561

If you look at the first places in each order, you find them single because they are tetragons and are set in an odd number place and because they are first. If you look at the third place you will see 4 and 9; here the former proceeds from two, and the number three creates the other. They are

established in an odd number place. If you look at the fifth place, you see
16 and 81; while one is born from four, the other is from nine. If you look
at the ninth place you will note the tetragons 256 and 6561 of which the
upper is 16 squared and the lower is 81 squared. If you do the same into infinity, there will be no change.

**39. That cubes participate in the same substance and are born of odd num-
bers.**

The cubes are, as it were, elevated by three intervals, on account of
equal multiplication, and participate in the same immutable substance and
are companions of the same nature and are produced by the grouping of
only odd numbers, never of even numbers. If all the odd numbers from
unity are put down, the joined figures will give the cubic numbers:

1	3	5	7	9	11	13	15	17	19	21

In these, the one which is first makes the first cube by power and
force. The two which follow, when put together, that is three and five,
create the second cube which is eight. The three which follow, joined together, that is seven, nine, and 11, create a cube which is contained in the
number 27, and this is the third cube. In the following four fours and the
following five fives according to this same manner, as many cubes are produced as odd numbers are put together by conjoinings. The following diagram illustrates it faithfully:

1	3,5	7,9,11	13,15,17,19	21,23,25,27,29
1	8	27	64	125

40. Concerning proportionalities.

There has been enough said about these things. Now a certain thing
occurs to us as we discuss proportions which can be useful concerning musical thought or astronomical subtleties, about the power of geometrical
speculation or merely the understanding of ancient mathematical writings,
and this will most conveniently terminate an introduction to arithmetic.[39]
Proportionality is the taking or collecting of two or three or whatever
number of ratios together. So we would in general define it: proportionality is a similar relationship of two or more ratios even if they are constitut-

---

39. Nicomachus, Book 2, chap. 21.

ed not by the same but by different quantities. The difference between numbers is a quantity. A proportion is a certain relationship to each other of two terms, as it were, that are contained on one concept, and that which their joining together will produce is a proportion.[40]

From joined ratios, an added proportionality emerges. In three terms the minimal proportionality is found. It occurs also in more terms, but it is greater. As two is to one and as these are two terms, they have a duplex proportion. Four compared to two is also a duplex proportion. Next, if you consider three terms, there is a proportionality of one to two and two to four. Here is a proportionality, as they say, and a collection or reduction of proportions to one. It is also found in larger numbers. If you add four to eight and these to 16 and to these 24 and more doubles, in those which follow there is a duplex proportionality in all the numbers emerging from double proportions. Therefore as often as one and the same term thus communicates with two terms around itself, to one it is a leader and to the other it is a follower. This is called a continuous proportion, as one, two, four. There is also a certain equality in these proportions; thus, as are 4 to 2, so are two to one, and again, as are 2 to one, so are 4 to two. This is also true according to the quantity of the number. By as much as three surpasses two, by so much does two surpass one, and by as much as one is less than two, by so much is two surpassed in three. But if another term is compared to one, or some other term to a different, it is necessary that this be called a disjunct relation, so that in terms of the numbers of the proportion, they are 1, 2, 4, 8. As two is to one, eight is to four; conversely, as one is to two, so four is to eight; and round about, as four is to one, so eight is to two. According to their quantity, the numbers are in a relation of this fashion: 1, 2, 3,4. By as much as one is surpassed by two, by so much is three surpassed by four; and by as much as two surpasses one, by

40. It is clear that at this point Boethius is not going to speak of a proportion between two terms, or what is normally called a *ratio*, but of a relation between three and, later, four elements. This inconsistency of terminology derives from an uncertainty in the terminology of Nicomachus. Boethius uses the terms *proportionalitas* and *medium* interchangeably here, as his Greek source did the comparable terms ἀναλογία and μέσοτης . Nesselmann, *Die Algebra der Griechen* (Berlin, G. Reimer, 1842), argued that ἀναλογία applied originally only to geometrical proportions and μέσοτης to arithmetic and harmonic proportions (p. 210). His argument is based on Iamblichus (Pistelli, pp. 100, 104) as well as on Nicomachus. But his point is not consistently verifiable. In any case, the context of Boethius usually makes his meaning clear. See also the note on the distinction between *proportio* and *ratio*, above, p. 26 and D'ooge, pp. 264-65.

so much does four surpass three. Rearranged, then, by as much as one is lower than three, by that much is two less than four; by as much as three surpasses unity, by that much does four surpass two.

41. What proportionality was among the ancients; what later thinkers added.

It is testified to and known among the ancients who have studied the learning of Pythagoras, or Plato, or Aristotle, that these are the three ways to knowledge: arithmetic, geometric, harmonic.[41] After these relationships of proportions there are three others, which are conveyed to us without names but are called fourth, fifth, and sixth, and which are contrasted with the above. Then later thinkers, on account of the perfection of the number ten, which was pleasing to Pythagoras, added four other kinds, so that in these proportionalities they brought together a body of proportions ten in number. According to this number we describe the prior relationships and comparisons where there are five in the major proportions, which we call leaders, and with them we put five others, minor terms, which we call followers. Also in Aristotle's and Archytas' description of the ten predicaments, the Pythagorean ten is manifestly found. So Plato, a very zealous student of Pythagoras, divided them according to the same argument, and Archytas the Pythagorean, before Aristotle, even though in some it may seem ambiguous, established those same ten predicaments.[42] So also there are ten parts among these groups and many other tens, and it is not necessary to pursue all of these here.

42. First we must speak of that which is called arithmetic proportionality.

Something must be said concerning proportionality and middle terms;[43] first we will treat of mediation which, according to the equality of

41. Nicomachus, Book 2, chap. 22. On the history of these proportions, see Imblichus, pp. 10, 116; Proclus, *Commentary on the Timaeus*, Book 2, chap. 18 and 29; D' ooge, p. 266.
42. The mystical significance of the number ten among the Pythagoreans was well known to the ancients. It symbolized the totality of universal existence and by the four numbers of the tetraktys ($1 + 2 + 3 + 4 = 10$) the Pythagoreans swore their most sacred oath. See *Joannis Stobaei Anthologium*, ed. Curtius Wachsmuth and Otto Hense (Berlin, Weidmann, 1884), Vol. 1, p. 16 and Aristotle, *Metaphysics*, Sec. 986 a. Archytas was long considered to author of a book on the categories. See D'ooge, p. 95.
43. Nicomachus, Book 2, chap. 23; Jordanus, Book 2, prop. 2; Book 2, prop. 16; Book 10, prop. 1, 5.

quantity, consists in the relationships of constituted terms, regardless of the equality of proportions. In these quantities, mediation is so used, and in these it must be seen where the terms differ from each other. What is meant by the differences of terms has been said above. Reason itself declares that arithmetic proportionality is involved in mediation of numbers since its proportion consists in the quantity of number. Now for what reason should the relation of these types of terms which are called arithmetic be put before all other proportionalities? First, because the nature of numbers and the natural force of quantity places them in the principal place for us. Immediately we can recognize proportions of this type which refer to the differences of terms, as it will later be demonstrated, in the natural disposition of number. Thus it is seen where I argued in the previous book that arithmetic is the force prior to geometry and music when it was proposed that it does not infer the others and that if taken away from the others it will logically destroy them. So this argument proceeds in order, if one must start with arithmetic when discussing medial numbers, which deals with the difference of number not entirely in terms of proportions.

43. Concerning the arithmetic medial proportion and its properties.

We call that an arithmetic medial proportion in which as often as in three or any number of stated terms there is found the same and equal difference between all the terms.[44] In an undisturbed equality of ratios in terms and of differences, this is the order preserved:

1	2	3	4	5	6	7	8	9	10

In this disposition of the natural number, if one takes care to look continually at differences of terms, there is an equal difference of terms according to an arithmetic interval. The differences are equal, but the relationship and ratio are not equal. If one considers three terms, a continual proportion may be remarked; yet one of these is the leader and another the follower. Here indeed are two terms in relation to a middle one, and this is called a disjunct medial proportion. If therefore you look at only three terms according to a continual proportion or to four or to however many others according to a disjunct proportion, you will always see the same differences of terms, with only the proportions changed. If someone notes it in one case, the consequent ratio will not elude him.

44. Nicomachus, Book 2, chap. 23; Theon, p. 187.

Let there be a continual medial proportion of 1, 2, 3. Here one stands from two, and two from three, by only one number, and they are the same differences, but the proportions are different. For two is a duplex to one, three is a sesquialter to two. You will see the same in the remaining numbers. If, however, you choose other numbers, mix them and put them one ahead of the other, then place them into this type of relationship, the same thing can happen. If you intersperse equal terms and they go before each other by a prior disposition, then you put a single one between, you will note a difference of only two; if you put two you will note differences of three, if you put three, there are differences of four, if four, then of five. And so according to this mode, this difference of terms which we seek will be one more than the one you just put in. So if in three terms single ones are omitted, two will always intervene:

Differences

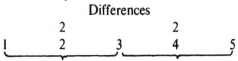

You see that since in the above disposition of the natural number the terms precede each other by single numbers, then with 2 and four left out, one to three and three to five, when compared, retain only two in their differences. This observation will be seen even in a disjunct order.

Differences

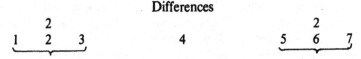

In such steps, no error will mislead one who insists on that similitude. Now if you put two between, three will contain the difference; if three, four will contain the difference; if four, five contain the difference, equally in continued and disjunct proportions. But the quality of the proportion will not be the same although the terms are distributed by equal differences. If they are put conversely, that same quality of proportion does not occur with the same differences; such a proportion is called geometric and not arithmetic.

It is characteristic of this proportion that there should be a possibility of three terms, so that with the extremes put together that sum which is between the extremes is not only in the middle place, but it is also the middle in quantity if the other two are added together. So if 1, 2, 3 are put down, one and 3 give four; two, which is the middle between them, is found to be half of four. If you multiply the middle by two it is equal to the extremes: twice 2 creates 4. But if it is a disjunct proportion, that sum which comes from two extremes put together is also brought about by the two medial terms. If there are numbers 1, 2, 3, 4, one and four make 5, two and 3 of the middle part also make five:

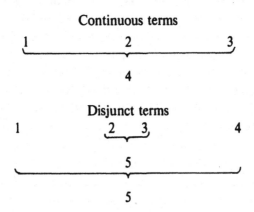

Continuous terms

Disjunct terms

To this there is added, as in an essential property, that just as all these terms are in arranged disposition in relation to each other, so are differences arranged in relation to differences. Thus, each term is equal to itself and the differences are equal to the differences.

There is also that more subtle matter which many learned in this discipline, except Nicomachus,[45] never observed before; that is, in every disposition, either conjunct or disjunct, the number which comes from the two extremities multiplied is less than that number which is made from the median multiplied by itself by as much as the two differences contain between themselves compared to the terms which are constituted between them when multiplied together. Let us put down three terms in his manner: 3, 5, 7. If the three is increased seven times, it comes to the number 21. If the middle term, 5, is multiplied by itself, then five times five makes 25. And this number is four larger than the one the extremities make when multiplied together. Between three and five, and five and seven, there are two, which multiplied together give 4. Two times two makes four. Rightly therefore is it said in this type of disposition that the extremi-

---

45. Boethius claims that Nicomachus originated this relationship. It appears again in Nicomachus' *Handbook of Music*, Book 3, chap. 8 (see D'ooge, p. 269). *Nicomachus* indicates it was his original observation by claiming that "it seems to have been *missed* by all."

ties multiplied are less than that number which comes from the middle squared by as much as the difference between the two numbers multiplied together.

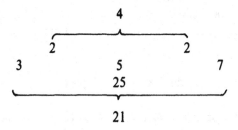

It may be noted that the fourth property of this kind of disposition, which the ancients considered most singular, is that in this medial proportionality there is a major proportion found in the smaller terms and a minor proportion in the larger terms. In this disposition of 1, 2, 3, the smaller terms are 1 and 2, the larger are 2 and 3. Two to one is a duplex, 3 to 2 is a sesquialter. But the proportion of a duplex is larger than a sesquialter. In an harmonic proportion it happens in a contrary way. In the smaller terms there are minor proportions; in the larger terms a major proportion is maintained. Between these two proportions, that is the arithmetic and the harmonic, the geometric proportion provides the middle position, which in either larger or smaller terms maintains equal quantities of numbers in proportionality. In the place between the larger and the smaller, the equality of the median is placed. Now enough has been said concerning the arithmetic medial proportion.

44. Concerning the geometric medial proportion and its properties.

In the chapter which follows, the geometric medial proportion is set straight,[46] which alone is most able to be called a proportionality since it is placed in the same ratio of terms whether we investigate it in its larger or its smaller terms. For in it an equal ratio is always kept and the quantity and multitude of number is regularly ignored, contrary to the arithmetic proportion. So are 1, 2, 4, 8, 16, 32, 64 or in triple proportion, 1, 3, 9, 27, 61 or in quadruple or quincuple or in any multiplication of numbers, when

46. Nicomachus, Book 2, chap. 24; Theon, p. 189; Jordanus, Book 2, prop. 1, 3, 5, 66; Book 9, prop. 38; Book 10, prop. 20.

a given extension of numbers is set up. In such numbers, whatever number of terms you take up, these will fulfill a geometric proportion in such a way that just as the prior is to the following, so is the following to the others; and again, if you exchange these it will be the same. If three terms are put down, 2 and 4 and 8, we see that just as 8 is to 4, so is 4 to 2. And if you reverse these, just as two is to four, so is four to eight.

2	4	8
duplex	duplex	

If you consider four terms, such as 2, 4, 8, 16, just as the first is to the third, that is 2 to 8, so will the second be to the fourth, that is 4 to 16. Each ratio is a quadruple. Conversely, as the fourth term is to the second, so the third is to the first. This is also valid disjunctly. As the first is to the second, that is two to 4, so the third is to the fourth, that is 8 to 16. Conversely, as the second is to the first, that is 4 to 2, so is the fourth to the third, that is 16 to 8. You will see the same in all the numbers with due consideration.

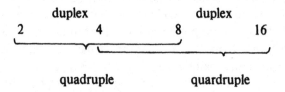

A medial proportion of this type has its own property, which is that in every disposition according to this proportionality of terms, the relationships of the differences are contrary to each other by the very terms between which the differences are found. Either the duplex terms are contrary to each other and the differences are duplex, or the terms are triple and the differences triple, or according to whatever multiplication, the multiplicity will be the same in the differences, as in the first instance it is found in the terms. So the following diagram shows:

Duplex differences

	1	2	4	8	16	32	64	128	
1	2	4	8	16	32	64	128	256

There can be no doubt that when all the terms are duplex, the differences between the terms will also be found duplex, so that with one term less in the differences, when all the terms are completely disposed beneath the terms of which they are the differences, the above order will emerge. There is also another property in that when every larger term is compared to a smaller, that term includes the difference from the smaller. Two differs from unity by a unity itself, and four differs from two by two, and eight differs from four by four, and so on; the other larger numbers differ from the smaller by the smaller number, and that is also the amount by which they surpass them. This also happens in the duplex proportion; if there is a triple proportion, the larger term differs from the smaller term by a number which is double of the smaller term, as in the terms 1, 3, 9, where three differs from one by two, to which unity, the smaller term reaches when it is doubled. Nine differs from 3 by 6, which the smaller number 3 will give when doubled. In all the other numbers, this type of rationale will be found. If there are quadruples, the larger term differs from the smaller by a triple of the smaller term. If they are quincuple, it is quadrupled; if sescuple, it is quincupled, and by one less multiplication does the comparison of smaller terms reach the larger terms and by it does a larger term surpass a smaller.

The smaller differences

	1	2	4	8	16	32	64	128	
1	2	4	8	16	32	64	128	256

Duplex terms

Double the smaller differences

	2	6	18	54	162	486	1458	
1	3	9	27	81	243	729	2187

Triple terms

Triple the smaller differences

	3	12	48	192	768	3072	12,288	
1	4	16	64	256	1024	4096	16,384

Quaduple terms

This proportionality is found in all the other superparticular or superpartient terms and is preserved in all of them by this property, so that in a continuous proportion whatever comes under the extremities, if they are three terms, it arises from a multiplied proportion. If we have 2, 4, 8, that which comes from twice eight is the same as that which comes from four four's; or if in four terms there is a disjunct proportion, whatever comes under each extremity, it grows from the multiplication of the two medians, so that if there are 2, 4, 8, 16, that which comes from twice 16 is also rendered by four 8's. This is a great and most certain example for us of the fact that, as we said, from equality are established all the species of inequality. In all the multiplex or superpartient or superparticular or other relations the geometric proportionality is kept, as we said above, and they contain all these properties. It is the fourth property of this relationship that in larger and in smaller terms there is always an equal proportion kept. If we put down the numbers 2, 4, 8, 16, 32, 64, among all these there is a duplex proportion. This proportionality would also appear in the double proportions from unity alternately in figures longer by one side and in squares disposed in the primary relationship of the multiplex from the duplex running through all the proportions and relations of the superparticular. This is designated in the diagram given below:

Tetragon	1	
Longer by one side	2	duplex
Tetragon	4	duplex
Longer by one side	5	sesquialter
Tetragon	9	sesquialter
Longer by one side	12	sesquitertian
Tetragon	16	sesquitertian
Longer by one side	20	sesquiquartan
Tetragon	25	sesquiquartan
Longer by one side	30	sesquiquinta
Tetragon	36	sesquiquinta
Longer by one side	42	sesquisexta
Tetragon	49	sesquisexta

45. Which medial proportions are compared to what things in the state of public affairs.

The arithmetic proportion may thus be compared to the state, which is ruled by a few, and so in the smaller terms there is a greater proportion. They say that a state of the very best is a musical proportion, because there is a proportionality in the larger terms. There is a geometric proportion when the state is of the people, as it were, and of a balanced citizenry. For in either larger or in smaller, the whole is put together with an equal proportionality of all, and there is an equality between all; there is a certain equal right balance in preserving proportions.

46. That plane numbers are joined by only one proportionality, but solid numbers are joined by two proportions placed medially.

After these matters, it is now time we relate something very useful in Platonic disputation which is treated in that all encompassing work, the *Timaeus*,[47] and which is by no means easy for anyone unless he is trained in a very discerning reason. All plane figures which do not grow by altitude are extended only by one kind of geometric medial proportion; no other may be found to keep it together. So, two relationships are constituted in such an interval, that is from the first term to the middle, from the middle to the third term. Now if they were cubes, they would have two middle numbers where a third relation would not be found according just to geometric proportion; hence solid forms are said to have three intervals. There is one interval from the first to the second, from the second to the third, from the third to the fourth, which is the most distant. Rightly therefore are plane figures said to be contained with two intervals and solid to be contained with three.

Let there be two tetragons, namely 4 and 9. One of these is able to be constituted the median in the same proportion. Now six to 4 is a sesquialter and 9 to six is also a sesquialter. Here it happens that each of the sides of the individual tetragons produces a median of six; the tetragon of four has a side of two, the tetragon of nine has a side of three. These multiplied give six: twice three is six. So as often as with two tetragons given we wish to find their middle term, their sides should be multiplied, and that which is produced from these is the median. If these are cubes, as 8 and 27, two such numbers are constituted with the same proportion of the median between them, namely 12 and 18. Now 12 to 8 and 18 to 12 and 27 to 18 are joined by a sesquialter proportion only. In these also there is the same ratio of sides. For from one cube, which is the closer, one median reaches two sides, each one by an alternately posed term. It is also the same in the other medial proportion. Let us put down two cubic numbers and between them the two median terms, which we have named above: 8, 12, 18, 27. The side of eight is binary; twice two gives eight. The side of the cubic

---

47. *Timaeus*, Sec. 32 A-B. Even with a most discerning reason *(penetrabili ratione)*, modern critics have had difficulty interpreting Plato's text. For discussions of Plato's meaning, see D'ooge, p. 272 and *The Timaeus of Plato*, Edited with introduction, translation, and notes, R.D. Archer-Hind (New York, Macmillan, 1888), pp. 97-99. Nicomachus himself confuses the meaning by misquoting Plato. Boethius must have given the matter some consideration since he acknowledges its difficulty and traces the question back to the *Timaeus*, which Nicomachus does not mention.

number 27 is ternary; three times three thrice gives 27. Thus, the median number which is the next to eight is 12. It is affected by the two sides next to it, from eight on one side and on the other by the cubic number 27. Two two's three times gives 12. Also 18 by the same logic is affected by the two numbers next to it, on the one side by the cubic number 27, on the other side by twice 8. Three threes twice gives 18.

This proportion must be universally observed. If tetragon multiplies tetragon, without doubt a tetragon will emerge; but if a figure longer by one side multiplies a tetragon or a tetragon multiplies a figure longer by one side, the figure longer by one side always grows from it. Again, if a cube multiplies a cube, the form of a cube is created; if a figure longer by one side multiplies a cube or a cube multiplies a figure longer by one side, a cube is never created. This occurs according to the similarities of even and odd. If even multiplies even, an even number is always born; if an odd multiplies an odd, an odd number is created. If odd multiplies even or if even multiplies odd, the even will always be overcome. This is understood more easily from the teaching of Plato in his book, *The Republic*;[48] there it is called the marriage number, which the philosopher introduces in the persons of the muses. Now we must go to the third type of medial proportion.

47. Concerning harmonic medial proportions and their properties.

An harmonic medial proportion[49] is one constituted by neither the same differences nor by equal proportions but it is one in which as the highest term is when compared to the smallest term, so the difference of the larger two is when compared to the difference of the middle term and the smallest. Such are 3, 4, 6 or 2, 3, 6; six exceeds four by a third of itself, and that is two; four surpasses three by a fourth of itself, and that is one; six exceeds three by half of itself, that is three; three exceeds two by a third of itself, that is by one. So in these there is not the same proportion of terms, nor are there the same differences, but as the larger term is when compared to the smaller term, so the difference between the middle and

---

48. *Republic*, Sec. 546.
49. See Nicomachus, Book 2, chap. 25; Jordanus, Book 3, prop. 32; Book 10, prop. 34, 36, 40. The three proportions, geometric, arithmetic, and harmonic, were originally developed by the Pythagoreans (Iamblichus, p. 100) but the third type was called ὑπεναντία. It was changed to ἁρμονικά (harmonic) by the schools of Archytas and Hippasus when they considered it in terms of musical relationships, and the two disciplines, arithmetic and music, came to be studied together. See also Theon, p. 189.

the largest is when compared to the difference between the middle and the one lower. In the proportion of 3, 4, 6, the larger term, which is 6, is a duplex to the smallest term, which is 3, and the difference between the largest and the median, that is of the six and 4, is two. The difference between the median and the last term, that is of four and three, is a unity, and it is seen as a duplex. Such is shown in the following diagram:

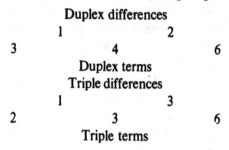

This medial proportion has a property, it is said, contrary to the arithmetic medial proportion. In the case of the arithmetic proportion, in its minor terms there is a major ratio; in the major there is a minor ratio. In the harmonic proportion, however, the ratio in the major terms is larger, and in the minor terms, it is smaller. In the disposition 3, 4, 6, three compared to four is a sesquitertian relationship, and six to four gives a sesquialter relation. But a sesquialter is a larger proportion than a sesquitertian and the larger transcends the middle term by as much as a third part. Therefore it is also rightly a kind of geometric median and properly is it considered such a proportionality because in an arithmetic proportion where in the larger term there is a minor ratio and in the minor there is a larger ratio, yet in a geometric, in the larger terms there is a larger ratio and in the minor terms there is a smaller ratio. That is truly a proportionality in which, while holding the place of a median between two other proportionalities, there are contained in both the major and the minor terms equal comparisons of ratios. And this is a sign that its median has a certain geometric proportion between extreme terms.

Now in an arithmetic proportion the middle term exceeds the minor to an amount of the same fraction of itself as the largest exceeds the middle. Let us have the arithmetic disposition 2, 3, 4. The third number exceeds two by a third part of itself, that is by one; in four this three is exceeded by the same part of itself, that is by one. But the number three does not surpass the smaller number by the same fraction of the smaller number, nor is it surpassed by a larger fraction of the larger. For it exceeds the smaller number, that is two, by one, which is half of that same two, and in four it is surpassed by one, which is a fourth of the four. Rightly is it said, therefore, that the middle term is in a medial proportion of this sort: by the same fraction of itself does it surpass the smaller term as it is surpassed by the larger, but not by the same fractions either of the smaller does it surpass the smaller, or of the larger is it surpassed by the larger.

The harmonic medial proportion has contrary properties. Not by the same fraction of itself does the median term in this proportion either surpass the smaller or is it surpassed by the larger, but by the same part of the smaller does it surpass the smaller by which fraction it is surpassed in the larger. In this harmonic disposition, which is 2, 3, 6, the number three surpasses the two by a third part of itself, and the same number three is surpassed by six in terms of its whole quantity, that is by three; the number three itself surpasses the smaller number by half of the smaller, that is by one, and it is also surpassed by half of the larger in the larger term, that is by three--three is half of six.

In a geometric medial proportion, the median term does not surpass the smaller by consistent amounts of itself, nor is it surpassed in the larger by consistent amounts nor by the same fraction of the smaller does it surpass the smaller, nor is it surpassed in the larger by a fraction of the larger. But by a certain fraction of itself does the median term surpass the smaller and by the same fraction of itself does the larger term surpass the median, which is as a middle, and one extremity by equal parts surpasses the median and the other extremity by its fraction. In the geometric disposition of 4, 6, 9, by a third part of itself does the median number six surpass the number four, that is by two; and again by a third part of itself does the number nine surpass the six, that is by three.

The harmonic medial proportion has another property, that when it adds the two extremities into one and multiplies it by the median, a duplex relationship is assembled when compared to the two extremities multiplied together. Put down these terms: 3, 4, 6. If you join the three and the six, you will make nine, which if multiplied by four brings about 36. If the extremities were multiplied by each other, and made three sixes, they would amount to 18, which is half of the prior sum:

	18	
	36	
3	4	6
	9	

48. Why it is called an harmonic medial proportion and how it is arranged.

Perhaps we should consider why we call this an harmonic medial proportion. The reason is that, since the arithmetic disposition divides equal quantities only according to differences, and the geometric joins terms through equal proportions, the harmonic proportion to something has related aspects of these because it has an attitude of proportion not only in its terms alone or in their differences alone, but commonly in both. It asks that just as there be extreme terms in a ratio to each other, so the difference of the larger to the smaller stands compared to the difference of the median to the final term. The relation to another thing is properly a consideration of harmonic proportion, as we saw in the division of all things mentioned in the first book. Of those musical consonances which they call symphonies, you will find practically all the ratios of the harmonic medial proportion. For that symphony called diatessaron, which is the principal one, and, as it were, the one holding the force of a primal element, it is constituted in an epitrita ratio, as a four is to three, and is found in harmonic medial proportions of such a type.

Let there be put down the terms of this type of harmonic medial proportion, of which the extreme terms are duplex, and another after it, of which the extremes are triple:

3	4	6
2	3	6

Six is the duplex of three in the first set, and six is the triple of two in the second set. If we put together the differences of these and compare them to each other, the epitrita proportion is assembled, from which resounds the symphony of the diatessaron.

Between three and six there is three, and between two and six there is four, which compared to each other bring about a sesquialter proportion.

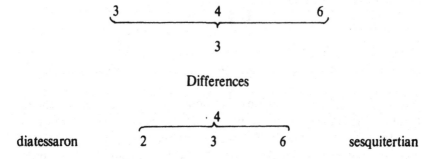

Differences

In the same kind of medial proportion is found the constitution of a diapente symphony which a sesquialter relationship will set up; of the sets which are put below, in two, the six to four is a sesquialter and in the third, there is a three to two. From both of these a diapente symphony is constructed.

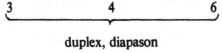

	3	4	6	sesquialter, diapente
sesquialter, diapente	2	3	6	

After this comes the consonance of the diapason, which is made from a duplex, as in the formula below:

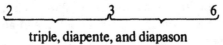

3   4   6

duplex, diapason

In the triple disposition, likewise the diapente and diapason symphony is put together, keeping the sesquialter and duplex ratio, which the diagram below shows:

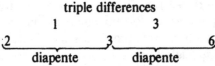

2   3   6

triple, diapente, and diapason

Since a triple contains two consonances, namely the diapente and diapason, in a disposition of this triple, we will find again the same triple in their differences, according to the mode described below:

triple differences

1   3

2   3   6

diapente   diapente

In a duplex disposition the larger term compared to the difference of the median term against it is a triple; and again the smaller term against the difference of the smaller term is a triple.

Differences

1   2

3   6

triple   4   triple

The larger symphony which is called a double diapason is, after a manner, twice two since the diapason symphony is put together from a duplex proportion and since to form it there is an harmonic joining at the middle. In a duplex proportion the middle term is found to be quadruple the amount of its minor difference:

$$
\left.\begin{array}{lll}
& 1 & \\
3 & & 4
\end{array}\right\} \quad \begin{array}{ll}
& \text{quadruple} \\
6 & \text{twice diapason}
\end{array}
$$

In triple extremities the larger difference to the minor difference is quadruple and produces a double diapason symphony. For, in the disposition 2, 3, 6 the difference between the extremes, that is between 2 and six, is four; the smaller differences, that is between three and two, is one. Four is a minor relation of a quadruple to one and this comparison contains the consonances of a double diapason.

49. Concerning geometric harmony.

An harmonic median is called such because its proportionality is related to geometric harmony.[50] They call a cube geometric harmony because it is so extended from longitude into latitude and it also grows into an accumulation of height so that starting out from equals and going to equals, it has developed so that it fits totally evenly in relation to itself. This type of medial proportion exists in all cubes which, it has been seen, is a geometric harmony.

Every cube has 12 sides, 8 angles, 6 surfaces. This order and disposition is harmonic. Let 6, 8, 12 be put down.[51] Here, as the larger term is to the smallest, so is the difference of the larger and the median compared to the difference of the median and the smallest. The terms 12 and 6 are duplex; the difference between twelve and eight is four, and between eight and six it is two. Four compared to two is a duplex ratio. Again, eight, which is the median, precedes the smaller term by a fraction of itself and by another fraction is it preceded in the larger term. By the same fraction of the minor does it surpass the minor, as is the fraction of the larger by which it is surpassed in the larger. Again, if the extremities be added together, and this sum is multiplied by the middle eight, that number will be a duplex to the total which the extremities make when multiplied together.

We will find all the musical symphonies in this disposition. A diatessaron is 8 to 6, since this is a sesquitertian proportion; the diapente is 12 to 8, which is called a sesquialter comparison and such a comparison contains a diapente. A diapason is born from a duplex produced from the

50. Nicomachus, Book 2, chap. 26.
51. See *De Institutione Musica*, Book 1, chap. 10 where Pythagoras is said to encounter four blacksmiths whose resounding hammers come to 6, 8, 9, 12 measures of weight and give off sounds at corresponding musical pitches.

adding together of 12 and 6. The diapason and the diapente, which maintain the ratio of a triple, come from the difference of the extremities compared to the smaller difference. The difference between twelve and six is 6, and the smaller difference between eight and six is 2. Six to two is a triple, and the diapason with the diapente sounds a consonant. The major consonance, that is a double diapason, comes from a quadruple, and is seen in a comparison of the middle term, that is eight, and the difference which is found between eight and six. Thus a proportionality of this type is properly and conveniently called a proportionality of an harmonic median.

50. How, in two terms set up opposite each other, are arithmetic, geometric, and harmonic medians exchanged among themselves; in what their generation consists.

We should make manifest how, when a reed with holes put into it is given to us,[52] such is the custom in playing music on this instrument that, leaving a middle hole open then variously opening and closing the other holes with one's fingers, the holes emit diverse sounds; or with two chords stretched next to each other, a musician raises the pitch by stretching, or lowers it by loosening, the string. Thus with two numbers given, we may introduce now the arithmetic, now the geometric, and now the harmonic medial proportion. Thus the right and proper name of the medial proportion may be given when, with the extremities remaining unchanged, the median may be seen to move to this or that number, and so be carried back and forth. We will also be able, within these two terms placed opposite each other, to change the median to either even or odd numbers so that, when we place an arithmetic median only, the ratio and equality of differences will be preserved; when we place a geometric median, the ratio produced keeps itself proportional, and if we insert an harmonic median, the comparison of the differences will not disagree with the proper proportion of its terms.

First let certain even extremities be posed, between which it will be necessary for all these medial terms to appear; these are 10 and 40. Then if between these I put 25, I will have an arithmetic proportion in a quantity of differences immutably preserved and in this kind of disposition: 10, 25, 40. You see that the terms exceed each other by the quantity of fifteen; all the properties which we said above occur in an arithmetic medial propor-

52. Nicomachus, Book 2, chap. 27.

tion you will find no different in this disposition. Just as each one of the terms is to itself, since each one is equal to itself, so are the differences to each other, that is, they are equal to each other. By so much as the larger term surpasses the middle, by that much also does the middle surpass the smaller. The total of the extreme terms is a duplex to the median; the proportion of the smaller terms is in such a comparison as exists between the larger terms. So much smaller is the number made from the extremities multiplied together from the number made by the median multiplied by itself as the differences multiplied will give. Note also that the middle number is surpassed in the larger by that fraction of itself that the median surpasses the smaller, but not by the same fraction of the smaller does the median surpass the smaller nor by that same part is the median short of the larger. All these properties are none other than of the arithmetic medial proportion which, if the reader remembers the things said above, he will undoubtedly recognize to be such.

Again, if between that 10 and 40 I put 20, immediately a geometric medial proportion arises, with all its properties, in place of the arithmetic medial proportion. In the disposition of 10, 20, 40, just as the largest is to the middle, so the middle is to the smallest. The number contained by the two extremes multiplied is equal to that reached in multiplying the middle by itself. The differences are in the same ratios as the three terms. There is no increase or diminution of proportions according to the terms; the proportions of the larger terms do not differ from the proportion of the smaller terms.

If I should wish to put together an harmonic medial proportion I would have to put the number 16 between each of the extremes so that they would be disposed in this manner: 10, 16, 40. Now it is permitted in this kind of disposition to acknowledge all the harmonic properties. By whatever ratio the largest number is joined to the smallest, by the same ratio are the differences compared to each other. By whatever fraction of the larger number that the middle number is surpassed in the larger number, by the same fraction of the minor does the middle surpass the minor. In the larger terms there is a major proportion; in the smaller terms there is a minor proportion. If the extremes are added together and multiplied by the medial term, that multiplication emerges in a duplex of what is created by the extremes alone multiplied.

All of this is established with even numbers. If odd numbers are put down, as 5 and 45, the apt middle term of 25 will constitute an arithmetic medial proportion. If we have the disposition of 5, 25, 45, each of these terms surpasses the other by the same quantity. Every property of the arithmetic medial proportion given above is preserved in these terms. But if I put the term 15 as the medial number, so that we have the disposition of 5, 15, 45, the terms are changed into a geometrical medial proportion, with equal ratios of the terms preserved between each. If I put 9 between each of the given terms, so that there is a disposition of 5, 9, 45, it becomes an harmonic medial proportion, so that by whatever sum the largest number surpasses the smallest, by the same ratio does the larger surpass the smaller difference.

Now we must explain by what method we are able to find a medial proportion of this sort. With two terms given it is necessary to establish an arithmetic medial term; each extreme term must be joined together and whatever comes from that joining should be divided and the number which results from that division is located between the extremities. This method creates an arithmetic medial term. So if I join 10 and 40, they produce 50, and if I divide that, 25 remains. This is the middle term according to an arithmetic proportion. If you take half of the number by which the larger surpasses the smaller and add it to the smaller, then the number that results you put in the median and an arithmetic median is formed. Now 40 is 30 more than ten and if you divide 30 in half you find 15. If you put this in the middle, the order of an arithmetic medial proportion is formed. If you would investigate a geometric ratio, find the tetragonal side of that number which is contained within both the extremities, and put this in the middle. Now under 40 and the number 10, 400 is contained. If you multiply ten by 40, this number results. Find therefore the side of this tetragonal number; it is 20. Twenty times 20 gives 400. With the squared side found, you establish the medial number. If you divide that proportion which the given terms keep between themselves, that which remains you put as the middle term. Now 40 is a quadruple to ten, so if you divide the quadruple, you produce a duplex, which is 20. Twenty is a duplex of ten. If you establish this as the median, it creates a geometric medial proportion. You would find the harmonic medial proportion by a similar method. Multiply the difference of terms by the minor term then join the terms and to the one which is thus made, put that number which is produced from the differences and from the minor term. When you have found the side of this term, you add it to the minor term, and the number which is made from that process, put as the middle term. Ten and 40 make 50. The difference between ten and 40 is 30 and if you multiply it by ten, that is by the minor term, you produce 300. Put this 300 next to the number which is made by the two terms joined together, that is next to 50. They make fifty times six, so the side of six is found. If you add this to the minor term, they make 16, and this number is constituted the median between ten and forty, and will serve as an harmonic medial proportion.

51. Concerning the three medial proportions which are contrary to harmonic and geometric proportions.

These then are the medial proportions[53] which we have treated at length and lucidly, which particularly are found in the writings of the ancients, and were discovered and tested by them. Their usefulness applies to almost every avenue of understanding. By-passing some other proportions, we have ignored those others since they are not much use to us in reading the works of the ancients, but only in fulfilling the quantity of the number ten. These we outline now lest they be out of reach or be unknown to some of you. It will be seen that these are contrary to the medial proportions given above from which they draw their origin, yet from them are these other constituted.

There is a fourth proportion which is opposite to the harmonic proportion. In this, with three terms posited, just as the larger term is to the smallest, so also is the difference of the smaller terms to the difference of the larger; so are 3, 5, 6. Six to three is in a duplex relationship. The smaller terms are five and three; the larger terms are 5 and 6. The difference between the smaller terms, that is 5 and 3, is 2; the difference between the larger terms, that is of five and six, is one, and if one compares two to one, he finds a duplex relation. Just as the largest term is to the smallest, so the difference of the smaller terms is to the difference of the larger. It is clear that this proportionality is opposed to and in a certain sense contrary to the harmonic proportion because there, just as the largest term is to the smallest, so the difference of the larger terms is to the difference of the smaller; but here it is just the opposite. It is also characteristic of this medial proportion that the product of the larger terms is double to that of the smaller terms. Six times 5 is 30, five times three is 15.

Two other medial proportions, that is the fifth and the sixth, are both contrary to the geometric medial proportion and are seen to be opposed to it. The fifth medial proportion occurs when as often in three terms just as the middle term is to the smaller term, so their difference is to the difference of the middle and larger. Now in the disposition of 2, 4, 5, four is a duplex to the two, and between the four and the two there are two. Between the four and the larger term, that is five, there is only one; two to

53. Nicomachus, Book 2, chap. 28; Jordanus, Book 2, prop. 7.

one is a duplex. It is contrary to the geometric medial proportion because in the geometric, just as the larger term is to the smallest, so the difference of the larger terms is to the difference of the smaller. But here it is in a contrary manner, and just as the terms of the smaller numbers are related to each other, so is the difference of the smaller terms when compared to the difference of the larger terms. It is also characteristic of this kind of disposition that the number which is contained in the larger term times the median is duplex to that which is obtained with the extremities multiplied. So, five times 4 is 20, five times two is 10, and 20 is a duplex to ten.

The sixth medial proportion occurs when the three terms are set up and just as the largest term is to the middle, so the difference of the smaller terms is to the difference of the larger terms. In a disposition such as 1, 4, 6, the largest term is a sesquialter to the middle; the difference between the larger terms, that is between four and six, is two; three and two compared bring about a sesquitertian relation of proportionality. In the same way this medial relation is contrary to the geometric as the fifth type is, because the proportion of difference from the smaller to the larger terms is reversed.

52. Concerning the four medial proportions which were later added to fulfill the number ten.

Such are the six medial proportions of which three lasted from Pythagoras to Plato and Aristotle. Three others followed later which we discussed previously, and to which they added with their commentaries. The following age, as we said, put down four more medial proportions to fulfill the quantity of ten which one will not find to that number in the books of the ancients. These we will now arrange as briefly as we can.

The first of these in order is the seventh medial proportion, and it is put together in this manner. In three given terms, just as the largest is to the smallest, so the difference of the largest and the smallest is to the difference of the smaller terms, as in this disposition: 6, 8, 9. Nine is a sesquialter to six, of which the difference is three; the difference of the minor terms, that is of eight and six, is two, which when compared to the above term makes a sesquialter proportion. The second of these four, that is eighth in the line of proportions, is one in which as often as in three terms just as are the extremes when compared to each other, so their difference is when compared to the difference of the larger terms; so are 6, 7, 9. Nine compared to 6 is a sesquialter. Their difference is a three, which compared to the difference of the larger terms--this is of seven and nine, whose difference is two--gives a sesquialter proportion. The third medial proportion among these four, that is the ninth in the total series, occurs when, with three terms having been posited, whatever proportion the medial term keeps in relation to the smallest, the smaller of the extremes retains it when compared to the difference of the minor terms; so are 4, 6, 7. Compared to 4, 6 is a sesquialter, and their difference is 2. The difference between seven and four is three and if we compare this to the above two, a sesquialter proportion is assembled.

The fourth proportion, which is tenth in the complete order, is thought to exist in three terms when, in such a proportion as the middle term is compared to the smallest, by the same proportion the difference of the extreme terms emerges; so are 3, 5, 8. Five is the middle term and it is superbipartient to the three; the difference of the extremes, namely of eight and three, is five, which when compared to the difference of the major terms, that is of 5 and 8 whose difference is three, is itself also found to be in a superbipartient relation.

53. An outline of the ten medial proportions.

Let us then outline all the medial proportions in order; they are of this rank and may thus be easily understood:[54]

Arithmetic	first	1	2	3
Geometric	second	1	2	4
Harmonic	third	3	4	6
Contrary to harmonic	fourth	3	5	6
Contrary to geometric	fifth	2	4	5
Contrary to geometric	sixth	1	4	6
First of four	seventh	6	8	9
Second of four	eighth	6	7	9
Third of four	ninth	4	6	7
Fourth of four	tenth	3	5	8

54. Concerning the greatest and most perfect symphony shown in three intervals.

It remains now to discuss the greatest and most perfect harmony which, constituted in three intervals, holds great strength in the modulation and tempering of music and in speculation on natural questions.[55]

54. For proportions one to six, see Theon, pp. 113-15.
55. Nicomachus, Book 2, chap. 29. Iamblichus reports that this proportion was known among the Babylonians and was adopted from them by Pythagoras (p. 118, Pistelli).

Nothing more perfect of this type can be found in a medial proportion which is produced in three intervals and determines the nature and substance of a most perfect body. In this fashion we have demonstrated that a cube is extended in three dimensions into full harmony. A relationship of this type may be found if, after two terms are established and these come from three intervals, length, width, and depth, then the two medial terms of this type are so established and formed in those three intervals. These are either produced equally from equals through equals, or from unequals to unequals, unequally, or from unequals equally to equals, or in whatever other way. Thus they maintain an harmonic proportion, but compared in another way they make an arithmetic proportion, nor is a geometric proportion lacking, but it too has its place among them. Now, in four terms, if the first were to the third just as the second is to the fourth, by reason of these proportions maintained, a geometric medial proportion would be defined, and that term which is contained under the extremities multiplied would be equal to that which is put together by the two median terms multiplied by each other. Again, if the large number in the four terms contains such a difference from that which is next to it as the one next to the largest has to the smallest, a proportion of this type is put into an arithmetic aspect and the combination of the extreme terms will be a duplex to its own median. If among four terms, the third surpasses the fourth term by an even part of that fourth term, and is surpassed in the first by an equal part of that first, a proportion and a median of this type is seen as harmonic. The term which is found in the addition of the extreme terms and in the multiplication with a medial term is twice the two extremes multiplied.

Let us have an example of this disposition in the following manner: 6, 8, 9, 12. There is no doubt that these numbers are all firm quantities. Six is born of once twice three, 12 from twice two three's, and the medial terms are eight, which is once two fours, and nine, which is one three three's. Therefore all the terms are related to each other and are understood in three-fold dimensions of intervals.

In these terms, the geometric proportionality is found if we compare 12 to 8 or 9 to 6. Each comparison is a sesquialter proportion and the term contained under each extremity is the same as that which is made from the medial terms. That which is made from twelve times six is equal to that which is made from eight times nine. This is a geometric proportion.

There is an arithmetic proportion if twelve is compared to nine and nine is compared to six; in each there is a difference of three and the extremities joined are duplex to the median. If you join six and 12, you make

18, which to nine, the middle term, is a duplex. In those terms, therefore, we have seen a geometric and an arithmetic medial proportion.

There is here also an harmonic medial proportion if we compare 12 to 8 and again 8 to 6. By the fraction of six that eight surpasses six, that is by one third, by that same fraction of 12 is eight surpassed. Four, the number by which twelve exceeds eight, is a third part of 12. And if you join together the extremities, that is 6 and 12, and multiply them by 8, it gives 144. If the extremities are multiplied with each other, that is 6 and 12, they make 72, to which the number 144 is a duplex. So in this way we may find all the musical consonances; eight compared to 6 and 9 to 12 give a sesquitertian proportion, which is the same as a diatessaron consonance. Six compared to nine and 8 compared to 12 give a sesquialter proportion and a diapente consonance. Twelve considered in relation to six gives a duplex proportion and this sounds a diapason interval. Eight considered in relation to 9, those two terms contrasting with each other put together an epogdous which in musical modulation is called a tone, and this is the common measure of all musical sounds. This is the smallest interval of all. From it is known how many tones' difference there is between the diatessaron and diapente, since there is the difference of only an epogdous between the sesquitertian and the sesquialter proportion.

We append a diagram of all this below:

Geometric

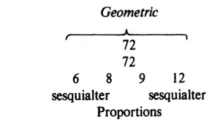

Arithmetic

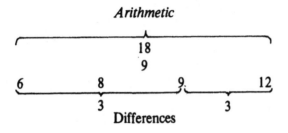

Harmonic

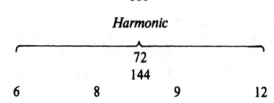

Fractions of the major and minor terms.

Consonances

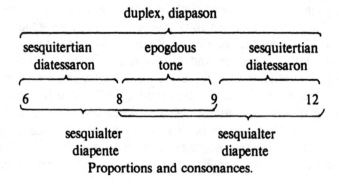

Proportions and consonances.

THE END

Bibliography

Aristotle. *The Basic Works of Aristotle,* ed. Richard McKeon (New York, Random House, 1941).

Bark, William. *Origins of the Medieval World* (Standford, Stanford University Press, 1958).
- »Theodoric vs Boethius: Vindication and Apology,» *American Historical Review,* 49 (1944), 410-26.

Beseler, H. and H. Roggenkamp. *Die Michaeliskirche in Hildesheim* (Berlin, G. Mann, 1954).

Bliss, A. J. ed. *Sir Orfeo,* (Oxford, Oxford University Press, 1966).

Beyse, Otto. *Hildesheim* (Berlin, Deutscher Kunstverlag, 1926).

Boethius. *Boetii De Institutione Arithmetica Libri Duo; De Institutione Musica Libri quinque; Accedit Geometria quae Fertur Boetii.* ed. Godofredus Friedlein (Leipzig, Teubner, 1867; reprinted 1966).
- »*Boethius« Geometria II: Ein Mathematisches Lehrbuch des Mittelalters* ed. Menso Folkerts (Wiesbaden, Franz Steiner Verlag, 1970)
- *The Consolation of Philosophy,* trans. Richard Green (New York, Bobbs-Merrill, 1962).
- *De Institutione Musica,* trans. Calvin Bower, (Yale University Press, forthcoming).

Boito, Camillo. *Il duomo di Milano e i disegni per la sua facciata* (Milan, Agnelli, 1900).

Bowie, Theodore, ed. *The Sketchbook of Villard de Honnecourt* (Bloomington, University of Indiana Press, 1955).

Bradwardine, Thomas. *Tractatus de Proportionibus.* ed. and trans. H. Lamar Crosby (Madison, University of Wisconsin Press, 1955).

Brommer, *F. Herakles: Die Zwölf Taten des Helden in Antiker Kunst und Literatur* (Münster, Manner, 1953).

Burnet, John. *Greek Philosophy* (London, Macmillan, 1914).

Butler, Basil Christopher. *Number Symbolism* (London, Barnes & Noble, 1970).

Cantor, Mortiz. *Vorlesungen über Geschichte der Mathematik.* 3 vols. (Leipzig, Teubner, 1892-98).

Chamberlain, David S. »Philosophy of Music in the *Consolatio* of Boethius,« *Speculum,* 45 (1970), 80-97.

Chaucer, Geoffrey. *The Works of Geoffrey Chaucer,* ed. F. N. Robinson. 2nd edition. (Boston, Houghton Mifflin, 1957).

Clemen, Paul. *Die Romanische Monumentmalerei in den Rheinlanden.* (Düsseldorf, Schwann, 1916).

Clerval, J. A. *L'enseignement des arts libéraux à Chartres et a Pàris* (Paris, A. Picard et fils, 1889).

Clichtoveus, Jodochus. *De Mystica Numerorum Significatione Opusculum* (Paris, Henricus Stephanus, 1513).

Conlee, John W. »The Meaning of Troilus' Ascension to the Eighth Sphere,« *The Chaucer Review,* 7 (1972), 27-36.

Cosman, Madeleine P. *The Education of the Hero in Arthurian Romance* (Chapel Hill, University of North Carolina Press, 1961).

Courcelle, Pierre. *La Consolation de Philosophie dans la tradition littéraire* (Paris, Etudes Augustiniennes, 1967).

Crombie, A. C. *Medieval and Early Modern Science* (New York, Doubleday, 1959).

Dante. *The Portable Dante,* trans. L. Binyon (New York, Viking Press, 1947).

D'Ancona, Paolo. »Le rappresentazioni allegoriche delle arti liberali nel Medievo e nel Renascimento,« *L'Arte,* 5 (1902), pp. 137-55, 211-28, 269-89, 370-82.

De Boissière, Claude. *L'Art d'arithmetique* (Paris, Annet Briere, 1554).

De Coussemaker, E. *Scriptorum de Musica Medii Aevi,* 4 vols. (Paris, A. Durand, 1864-76).

Delfino, Domenico. *Sommario di Tutte le Scienze* (Venice, Gabriel Giolito 1565).

Debelius, Franz. *Die Bernwardstur zu Hildesheim* (Strassburg, Heitz, 1907).

Dickson, L. E. *History of the Theory of Numbers* (Washington, Carnegie Institution of Washington, 1919).

Di Gnassi, Sylvestro. *Opera* (Venice, 1535).

Diophantus. *De Polygonis Numeris.* ed. Paulus Tannery (Leipzig, Teubner, 1893).

Diogenes Laertius. *Lives of the Eminent Philosophers.* ed. and trans. R. D. Hicks. 2 vols (New York, G. P. Putnam's Sons, 1925).

Dronke, Peter. »L'Amor che move il sole e l'altre stelle,« *Studi Medievali,* 3rd. ser., pt. 1 (1965), pp. 389-422.

Euclid. *The Thirteen Books of Euclid's Elements,* trans from the text of I. L. Heiberg by Thomas L. Heath. 3 vols. (Cambridge, Cambridge University Press, 1926; reissue 1956).

Fernelius, Johannes. *Quadratum Sapientiae* (Augsburg, Auguste Vindelicorum 1515).

Foeniseca, Johannes. *Quadratum Sapientiae* (Augsburg, Auguste Vindelicorum, 1515).

Fowler, Alaister, ed. *Essays in Numerological Analysis* (London, Routledge & K. Pual, 1970).

Frankl, Paul. *The Gothic: Literary Sources and Interpretations* (Princeton, Princeton University Press, 1960).

Friedman, John Block. *Orpheus in the Middle Ages* (Cambridge, Mass., Harvard University Press, 1970).

Gaffurio, Franchino. *Practica Musica,* ed. and trans. Clement A. Miller (Dallas, American Institute of Musicology, 1968).

Gerbert, Martin. *Scriptores Ecclesiastici de Musica Potissimum.* 3 vols. (San Blasianis, Typis San-Blasianis, 1784; reprinted, Rochester, University of Rochester Press, 1955).

Gilson, E. *History of Christian Philosophy in the Middle Ages* (New York, Random House, 1955).

Glarean, Heinrich. *De Vi Arithmeticae Practicae Speciebus* (Paris, Jacobus Gazellus, 1543).

Hambidge, Jay. *Dynamic Symmatry: The Greek Vase* (New Haven, Yale University Press, 1920).

Hanloser, H. R. *Villard de Honnecourt* (Vienna, 1935).

Haureau, B. *Notes et extraits de quelque manuscrits latins de la Bibliothèque Nationale* (Paris, Impr. Nationale, 1890-93).

Heath, Sir Thomas. *A History of Greek Mathematics* (Oxford, Clarendon Press, 1921).

Hesiod, *Works,* ed. Alois Rzach (Leipzig, Teubner, 1912); trans. into English by C. A. Elton (London, A. J. Valpy, 1823).

Hopper, Vincent Foster. *Medieval Number Symbolism* (New York, Cooper Square Publishers, 1938).

Horsley, Imogene. »Improvised Embellishment,« *Journal of the American Musicological Society,* 4 (1951), 5-20.

Hrosvitha. Opera, ed. Paulus Winterfeld (Berlin, Weidmannn, 1902).
- *Opera,* ed. H. Homeyer (Munich, F. Schoningh, 1970).

Iamblichus. *In Nicomachi Arithmeticam Introductionem Liber,* ed. Hermengildus Pistelli (Leipzig, Teubner, 1894).

Jansen, W. »Der Kommentar des Clarenbaldus von Arras zu Boethius De Trinitate,« *Breslauer Studien zu Historischen Theologie,* 8 (1926), pp. 12, 62, 108, 125.

Johannis de Muris, *De Arithmetica* (Vienna, 1515).

Katzenellenbogen, Adolph. »The Representation of the Liberal Arts,« *Twelfth Century Europe and the Foundations of Modern Society,* ed. M. Clagett, Gaines Post, R. Reynolds (Madison, University of Wisconsin Press, 18961), pp. 39-55,
- *The Sculptural Programs of Chartres Cathedral* (Baltimore, Johns Hopkins Press, 1959).

Klein, Jacob. *Greek Mathematical Thought and the Origin of Algebra,* trans. Eva Brann (Cambridge, Mass., M. I. T. Press, 1968).

Klingeberg, Heinz. »Zum Grundris der Ahd. Evangeliendichtung Otfrids,« *Zeitschrift für deutsches Altertum und deutsche Literatur,* Bd. IC, Heft 1 (1970), 34-45.

Kunstle, K. *Iconographie der Christlichen Kunst,* I (Freiburg, Herder & Co. 1908), 145-156.

Lax, Gaspar. *Proportiones* (Paris, Hermendus le Feure, 1515).

Macrobius, *Opera Quae Supersunt,* ed. L. van Jan 2 vols. (Leipzig, 1848-52).
- *Commentary on the Dream of Scipio,* trans. W. H. Stahl (New York, Columbia University Press, 1952).

Male, E. »Les arts libéraux dans la statuaire des moyen-âge,« *Revue Archeologique,* 17 (1891), 335-48.

Martianus Capella. *De Nuptiis Philologiae et Mercurii,* ed. A. B. Dick (Stuttgart, 1924; revised, Jean Préaux, Leipzig, Teubner, 1969). Trans. W.H. Stahl, New York (Columbia U. Press, 1975).

Masi, Michael. »Boethius and the Iconography of the Liberal Arts,« *Latomus,* 33, (1974), 57-75.
    - »The Christian Music of *Sir Orfeo,*« *Classical Folia,* 20 (1974), 3-20.
    -»Manuscripts Containing the *De Musica* of Boethius,« *Manuscripta,* vol. 15 (1971), pp. 88-97.
    - »A Newberry Diagram of the Liberal Arts,« *Gesta,* 13/1 (1973), 52-56.
    - »Recent Editions in Medieval Science,« *Cithara,* 13/1 (1973), 64-82.
    -»Troilus: A Medieval Psychoanalysis,« *Annuale Medievale,* 11 (1970), 81-88.

Mathenick, J. »De Boethii Morte,« *Eunomia,* 4 (1960), 26-37.

McNeill, George, P., ed. *Sir Tristrem* (Edinburgh, Blackwood and Sons, 1886).

Merzario, Guiseppe. *I Maestri Comacini: Storia artistica di mille duecento anni (600-1800),* (Milan, G. Agnelli, 1893).

Murdoch, John E. »Mathesis in philosophiam scholosticam introducta,« *Arts libéraux et philosophie au moyen-âge* (Montreal, Institut des études medievales, 1969), pp. 215-55.

Nemorarius, Jordanus. *De Arithmetica* (Paris, Joannem Stephanum, 1496).

Nesselmann, Georg Heinrich. *Die Algebra der Griechen* (Berlin. G. Reimer, 1842).

Nicomachus of Gerasa. *Handbook of Music,* in *Musici Scriptores Graeci,* ed. Richard Hoche (Leipzig, Teubner, 1866), pp. 235-82.
    - *Introduction to Arithmetic,* trans. Martin Luther D'ooge, intro. Frank E. Robbins and Louis Karpinski (New York, Macmillan, 1926).

Obrecht, Jacob. *Missa Maria Zart, Missae VII,* ed. M. Van Crevel (Amsterdam, G. Alsbach, 1960).

Paciuolo, Luca. *Summa de Arithmetica* (Toscalano, 1523).

Panofsky, E. »Die Entwicklung der Proportionslehre also Abbild der Stilentwicklung,« *Monatshefte für Kunstwissenschaft,* 14 (1921), 188-219.
  - *Meaning in the Visual Arts* (New York, Doubleday, 1955).

Parent, J. M. *La doctrine de la création dans l'école de Chartres* (Paris, J. Vrin, 1938).

Patch, H. R. *The Tradition of Boethius* (New York, Oxford University Press, 1935).

Philo Judaeus. *Works,* ed. and trans. F. H. Colson and G. H. Whitaker. 10 vols. (New York, G. P. Putnam's Sons, 1929-62).

Plato. *The Collected Dialogues,* ed. Edith Hamilton and Huntington Cairns (New York, Pantheon Books, 1961).
  - *Plato's Cosmology: The Timaeus of Plato,* trans. and commentary by Francis M. Cornford (New York, K. Paul, Trench, Trübner & Co., 1937).
  - *The Timeaus of Plato,* ed. and trans. R. D. Archer-Hind (New York, Macmillan, 1888).

Proclus. *In Primum Euclidis Elementorum Comentarii,* ed. G. Friedlein (Leipzig, Teubner, 1873).

Recorde, Robert. *The Grounde of Arts* (London, Reginald Wolfe, 1558).

Reese, Gustave. *Music in the Renaissance* (New York. W. W. Norton, 1959).

Regius, Hudalrich. *Utriusque Arithmeticae Epitome* (Freiburg, Stephanus Grauius, 1550).

Ringelbergius, Joachim Fortius. *Opera* (Leiden, Gryphium Lugdinum, 1531).

Ruffus, Gerardus, ed. and commentary. *Boethii De Arithmetica* (Paris, Simon Colinaeus, 1521).

Sachs, Curt. *Rhythm and Tempo: A Study in Music History* (New York, Norton, 1953).

Shelby, Lon. »The Education of the Medieval English Master Masons,« *Mediaeval Studies,* 32 (1970), 1-26.
   - »The Geometrical Knowledge of the Mediaeval Master Masons,« *Speculum,* 47 (1972), 395-421.

Smith, David Eugene. *Rara Arithmetica* (Boston, Ginn and Co., 1908).

Stapulensis, Faber. *Epitome Boethii* (Paris, Wolphganus Hopilius et Henricus Stephanus, 1503).

Stobaeus, Joannes. *Joannis Stobaei Anthologium,* ed. Curtius Wachsmuth and Otto Hense, 5 vols. (Berlin, Weidemann, 1884-1912).

Taylor, Thomas. *Theoretic Arithmetic* (London, A. J. Valpy, 1816).

Theon of Smyrna. *Exposition des connaisances mathématiques utiles pour la lecture de Platon,* ed. and trans. J. Dupuis (Paris, 1892; reprinted, Bruxelles, Culture et Civilisation, 1966).

Unicornus, Josephus. *De Utilitate Mathematicarum* (Venice, Dominicus de Nicolinis, 1561).

Valla Georgius. *De Arithmetica* (Venice, Aldus Romanus, 1501).

Van den Borren, Charles. *Guillaume Dufay: son importance dans l'évolution de la musique au XVe siècle* (Bruxelles, Maurice Lamertin, 1926).

Verdier, Philippe. »L'iconographie des arts libéraux dans l'art du moyen-âge jusqu'à fin du quinzieme siècle,« *Arts libéraux et philosophie au moyen-âge* (Montreal, Institut des études médiévales, 1969), pp. 305-355.

Von Simson, Otto. *The Gothic Cathedral.* (New York, Harper and Row, 2nd edition reprinted, 1964).

Wagner, Peter. *Geschichte der Messe, I Teil, bis 1600* (Leipzig, 1913).